Methods in Molecular Biology

Series Editor
John M. Walker
School of Life and Medical Sciences
University of Hertfordshire
Hatfield, Hertfordshire, AL10 9AB, UK

For further volumes:
http://www.springer.com/series/7651

mRNA Decay

Methods and Protocols

Edited by

Shireen R. Lamandé

Murdoch Children's Research Institute, The Royal Children's Hospital, Parkville, VIC, Australia

Humana Press

Editor
Shireen R. Lamandé
Murdoch Children's Research Institute
The Royal Children's Hospital
Parkville, VIC, Australia

ISSN 1064-3745 ISSN 1940-6029 (electronic)
Methods in Molecular Biology
ISBN 978-1-4939-7539-6 ISBN 978-1-4939-7540-2 (eBook)
https://doi.org/10.1007/978-1-4939-7540-2

Library of Congress Control Number: 2017961813

Printed on acid-free paper

This Humana Press imprint is published by Springer Nature
The registered company is Springer Science+Business Media, LLC
The registered company address is: 233 Spring Street, New York, NY 10013, U.S.A.

Preface

Cellular mRNA levels are determined by the balance between gene transcription and mRNA degradation. Modulating mRNA decay thus plays a vital role in post-transcriptional gene regulation and allows rapid changes in gene expression in response to developmental, metabolic, and environmental signals. Huge advances in understanding the global importance of mRNA decay have emanated from the introduction of genome-wide methods: microarrays and most recently RNAseq. Genome-wide sequencing has also been crucial in identifying mRNA sequence elements, the proteins that bind and target them for rapid degradation, and the large gene networks regulated by miRNA-directed mRNA decay. Nonsense-mediated mRNA decay is an important quality control surveillance mechanism that detects miss-spliced mRNAs with codon reading frameshifts and premature stop codons, or those with premature stop codons that result from genome mutations.

This volume brings together some of the latest methods that can be used to probe mRNA decay pathways and identify mRNA-binding protein targets as well as miRNA targets. The approaches will be important for molecular biologists, geneticists, and developmental biologists with an interest in understanding how normal development and tissue homeostasis are regulated and how these processes are perturbed in inherited and acquired diseases. Chapters 1–3 detail metabolic labeling and RNAseq methods for determining RNA decay rates. A method that can detect changes in the abundance of any mRNA in individual cells is described in Chap. 4. Different approaches for discovering RNA-binding protein targets are outlined in Chaps. 5–7. Integrating the large data sets generated by genome-wide sequencing approaches is a major bioinformatics challenge that is addressed by the methods included in Chap. 8. Chapters 9–11 focus on ways to identify miRNA targets and novel components of the miRNA-directed decay pathway, and Chaps. 12–16 describe a variety of recently developed approaches for studying nonsense-mediated mRNA decay.

I thank the authors, international experts with diverse expertise, for taking the time to write their chapters and for their patience and rapid responses during the editorial process.

Parkville, VIC, Australia *Shireen R. Lamandé*

Contents

Preface .... *v*
Contributors.......... *ix*

1 5′-Bromouridine IP Chase (BRIC)-Seq to Determine RNA Half-Lives 1
Toshimichi Yamada, Naoto Imamachi, Rena Onoguchi-Mizutani, Katsutoshi Imamura, Yutaka Suzuki, and Nobuyoshi Akimitsu

2 Determining mRNA Decay Rates Using RNA Approach to Equilibrium Sequencing (RATE-Seq) 15
Farah Abdul-Rahman and David Gresham

3 Metabolic Labeling of Newly Synthesized RNA with 4sU to in Parallel Assess RNA Transcription and Decay 25
Wei Sun and Wei Chen

4 Measuring mRNA Decay in Budding Yeast Using Single Molecule FISH 35
Tatjana Trcek, Samir Rahman, and Daniel Zenklusen

5 PAR-CLIP for Discovering Target Sites of RNA-Binding Proteins 55
Aitor Garzia, Pavel Morozov, Marcin Sajek, Cindy Meyer, and Thomas Tuschl

6 Characterizing mRNA Sequence Motifs in the 3′-UTR Using GFP Reporter Constructs 77
Rene Geissler and Andrew Grimson

7 iCLIP of the PIWI Protein Aubergine in *Drosophila* Embryos 89
Bridlin Barckmann, Jérémy Dufourt, and Martine Simonelig

8 Integration of ENCODE RNAseq and eCLIP Data Sets 111
Jorge Boucas

9 Identifying miRNA Targets Using AGO-RIPseq 131
Rebecca Petri and Johan Jakobsson

10 Integrated Analysis of miRNA and mRNA Expression Profiles to Identify miRNA Targets 141
Zhiming Li and Chi-Meng Tzeng

11 Identifying RISC Components Using Ago2 Immunoprecipitation and Mass Spectrometry 149
Tingfang Yi

12 Using Tet-Off Cells and RNAi Knockdown to Assay mRNA Decay 161
Thomas D. Baird and J. Robert Hogg

13 Identifying Cellular Nonsense-Mediated mRNA Decay (NMD) Targets: Immunoprecipitation of Phosphorylated UPF1 Followed by RNA Sequencing (p-UPF1 RIP–Seq) 175
Tatsuaki Kurosaki, Mainul Hoque, and Lynne E. Maquat

14 Generation of Cell Lines Stably Expressing a Fluorescent Reporter of Nonsense-Mediated mRNA Decay Activity 187
Nadezhda M. Markina, Anton P. Pereverzev, Dmitry B. Staroverov, Konstantin A. Lukyanov, and Nadya G. Gurskaya

15 Reactivation Assay to Identify Direct Targets of the Nonsense-Mediated mRNA Decay Pathway in *Drosophila* 205
Jonathan O. Nelson and Mark M. Metzstein

16 Studying Nonsense-Mediated mRNA Decay in Mammalian Cells Using a Multicolored Bioluminescence-Based Reporter System............... 213
Andrew Nickless and Zhongsheng You

Index *225*

Contributors

FARAH ABDUL-RAHMAN • *Centre for Genomics and Systems Biology, Department of Biology, New York University, New York, NY, USA*
NOBUYOSHI AKIMITSU • *Isotope Science Center, The University of Tokyo, Tokyo, Japan*
THOMAS D. BAIRD • *Biochemistry and Biophysics Center, National Heart, Lung, and Blood Institute, National Institutes of Health, Bethesda, MD, USA*
BRIDLIN BARCKMANN • *mRNA Regulation and Development, Institute of Human Genetics, UMR9002 CNRS-Université de Montpellier, Montpellier, Cedex 5, France*
JORGE BOUCAS • *Bioinformatics Core Facility, Max Planck Institute for Biology of Ageing, Cologne, Germany*
WEI CHEN • *Department of Biology, Southern University of Science and Technology, Shenzhen City, Nanshan District, China; Medi-X Institute, SUSTech Academy for Advanced Interdisciplinary Studies, Southern University of Science and Technology, Shenzhen City, Nanshan District, China*
JÉRÉMY DUFOURT • *mRNA Regulation and Development, Institute of Human Genetics, UMR9002 CNRS-Université de Montpellier, Montpellier, Cedex 5, France*
AITOR GARZIA • *Laboratory of RNA Molecular Biology, Howard Hughes Medical Institute, The Rockefeller University, New York, NY, USA*
RENE GEISSLER • *Department of Molecular Biology and Genetics, Cornell University, Ithaca, NY, USA*
DAVID GRESHAM • *Centre for Genomics and Systems Biology, Department of Biology, New York University, New York, NY, USA*
ANDREW GRIMSON • *Department of Molecular Biology and Genetics, Cornell University, Ithaca, NY, USA*
NADYA G. GURSKAYA • *Shemyakin-Ovchinnikov Institute of Bioorganic Chemistry, Moscow, Russia; Nizhny Novgorod State Medical Academy, Nizhny Novgorod, Russia*
J. ROBERT HOGG • *Biochemistry and Biophysics Center, National Heart, Lung, and Blood Institute, National Institutes of Health, Bethesda, MD, USA*
MAINUL HOQUE • *Genomics Center, Department of Microbiology, Biochemistry and Molecular Genetics, Rutgers New Jersey Medical School, Newark, NJ, USA*
NAOTO IMAMACHI • *Isotope Science Center, The University of Tokyo, Tokyo, Japan*
KATSUTOSHI IMAMURA • *Department of Microbiology and Molecular Genetics, Graduate School of Pharmaceutical Sciences, Chiba University, Chiba, Japan*
JOHAN JAKOBSSON • *Laboratory of Molecular Neurogenetics, Department of Experimental Medical Science, Wallenberg Neuroscience Center and Lund Stem Cell Center, Lund University, Lund, Sweden*
TATSUAKI KUROSAKI • *Department of Biochemistry and Biophysics, School of Medicine and Dentistry, University of Rochester, Rochester, NY, USA; Center for RNA Biology, University of Rochester, Rochester, NY, USA*
ZHIMING LI • *Translational Medicine Research Center, School of Pharmaceutical Sciences, Xiamen University, Xiamen, Fujian, China; Key Laboratory for Cancer T-Cell*

Theranostics and Clinical Translation (CTCTCT), Xiamen, Fujian, China; INNOVA Clinics and TRANSLA Health Group University, Xiamen, Fujian, China
KONSTANTIN A. LUKYANOV • *Shemyakin-Ovchinnikov Institute of Bioorganic Chemistry, Moscow, Russia; Nizhny Novgorod State Medical Academy, Nizhny Novgorod, Russia*
LYNNE E. MAQUAT • *Department of Biochemistry and Biophysics, School of Medicine and Dentistry, University of Rochester, Rochester, NY, USA; Center for RNA Biology, University of Rochester, Rochester, NY, USA*
NADEZHDA M. MARKINA • *Shemyakin-Ovchinnikov Institute of Bioorganic Chemistry, Moscow, Russia*
MARK M. METZSTEIN • *Department of Human Genetics, University of Utah, Salt Lake City, UT, USA*
CINDY MEYER • *Laboratory of RNA Molecular Biology, Howard Hughes Medical Institute, The Rockefeller University, New York, NY, USA*
PAVEL MOROZOV • *Laboratory of RNA Molecular Biology, Howard Hughes Medical Institute, The Rockefeller University, New York, NY, USA*
JONATHAN O. NELSON • *Department of Human Genetics, University of Utah, Salt Lake City, UT, USA; Life Sciences Institute, University of Michigan, Ann Arbor, MI, USA*
ANDREW NICKLESS • *Department of Cell Biology and Physiology, Washington University School of Medicine, St. Louis, MO, USA*
RENA ONOGUCHI-MIZUTANI • *Isotope Science Center, The University of Tokyo, Tokyo, Japan*
ANTON P. PEREVERZEV • *Shemyakin-Ovchinnikov Institute of Bioorganic Chemistry, Moscow, Russia*
REBECCA PETRI • *Laboratory of Molecular Neurogenetics, Department of Experimental Medical Science, Wallenberg Neuroscience Center and Lund Stem Cell Center, Lund University, Lund, Sweden*
SAMIR RAHMAN • *Département de Biochimie et Médecine Moléculaire, Université de Montréal, QC, Canada*
MARCIN SAJEK • *Laboratory of RNA Molecular Biology, Howard Hughes Medical Institute, The Rockefeller University, New York, NY, USA*
MARTINE SIMONELIG • *mRNA Regulation and Development, Institute of Human Genetics, UMR9002 CNRS-Université de Montpellier, Montpellier, Cedex 5, France*
DMITRY B. STAROVEROV • *Shemyakin-Ovchinnikov Institute of Bioorganic Chemistry, Moscow, Russia*
WEI SUN • *Department of Biology, Southern University of Science and Technology, Shenzhen City, China; Laboratory for Functional and Medical Genomics, Berlin Institute for Medical Systems Biology, Max-Delbrück-Centrum für Molekulare Medizin, Berlin, Germany*
YUTAKA SUZUKI • *Department of Medical Genome Sciences, Graduate School of Frontier Sciences, The University of Tokyo, Chiba, Japan; Department of Computational Biology, Graduate School of Frontier Sciences, The University of Tokyo, Chiba, Japan*
TATJANA TRCEK • *Department of Cell Biology, Skirball Institute of Biomolecular Medicine, NYU School of Medicine, New York, NY, USA*
THOMAS TUSCHL • *Laboratory of RNA Molecular Biology, Howard Hughes Medical Institute, The Rockefeller University, New York, NY, USA*
CHI-MENG TZENG • *Translational Medicine Research Center, School of Pharmaceutical Sciences, Xiamen University, Xiamen, Fujian, China; Key Laboratory for Cancer T-Cell Theranostics and Clinical Translation (CTCTCT), Xiamen, Fujian, China; INNOVA Clinics and TRANSLA Health Group University, Xiamen, Fujian, China*

TOSHIMICHI YAMADA • *Isotope Science Center, The University of Tokyo, Tokyo, Japan*
TINGFANG YI • *Department of Biological Chemistry and Molecular Pharmacology, Harvard Medical School, Boston, MA, USA*
ZHONGSHENG YOU • *Department of Cell Biology and Physiology, Washington University School of Medicine, St. Louis, MO, USA*
DANIEL ZENKLUSEN • *Département de Biochimie et Médecine Moléculaire, Université de Montréal, QC, Canada*

Chapter 1

5′-Bromouridine IP Chase (BRIC)-Seq to Determine RNA Half-Lives

Toshimichi Yamada, Naoto Imamachi, Rena Onoguchi-Mizutani, Katsutoshi Imamura, Yutaka Suzuki, and Nobuyoshi Akimitsu

Abstract

Analysis of RNA stability at genome-wide level is an advanced method in RNA biology that examines the half-life of each transcript. In particular, a pulse-labeling method using uridine analogs enables the determination of half-life of each transcript under physiologically undisturbed conditions. The technique involves pulse labeling of endogenous RNAs in mammalian cells with 5′-bromouridine (BrU), followed by measuring the chronological decrease of BrU-labeled RNAs using deep sequencing (BRIC-seq). Here, we describe a detailed protocol and technical tips for BRIC-seq.

Key words RNA degradation, RNA half-life, 5′-Bromouridine, Deep sequencing, BRIC-seq

1 Introduction

The abundance of individual mRNA transcripts in cells is determined by their transcription rate and their rate of degradation. Dynamic control of mRNA decay rate is involved in various cell signaling cascades, including responses to hormones [1, 2] or viral infections [3, 4] and cell cycle progression [5, 6]. Characterization of transcript degradation at the genome-wide level is an important step toward understanding the regulatory circuits for gene regulation.

Conventionally, genome-wide analyses of RNA degradation have used transcriptional inhibitors to measure ongoing decreases in RNA expression levels over time after shutting off transcription [7–11]. However, this inhibitor-based transcriptional arrest has a profoundly disruptive impact on the cellular physiology. In this chapter, we introduce an alternative strategy to monitor the degradation of transcripts: a pulse-labeling assay using modified

Toshimichi Yamada and Naoto Imamachi contributed equally to this work.

Shireen R. Lamandé (ed.), *mRNA Decay: Methods and Protocols*, Methods in Molecular Biology, vol. 1720, https://doi.org/10.1007/978-1-4939-7540-2_1, © Springer Science+Business Media, LLC 2018

nucleotides. Although there are several nucleotide analogs that can be applied in pulse-labeling assays [12–14], 5-bromouridine (BrU) is the suitable analog for labeling endogenous RNAs to monitor RNA degradation. BrU exhibits little toxicity toward cells and does not cause disincorporation of nucleotides during reverse transcription. By using BrU, we developed a novel method, designated BrU immunoprecipitation chase assay (BRIC) or BRIC through deep sequencing (BRIC-seq) [15, 16]. This method labels endogenous transcripts with BrU by its addition to the cell culture medium for a defined period. Total RNAs containing BrU-labeled mRNAs are isolated from the cells at sequential time points. The BrU-labeled mRNAs are then purified by immunopurification, and quantified by real-time polymerase chain reaction (qPCR) or deep sequencing (Fig. 1). By using BRIC-seq, we found that the mRNAs with long half-lives ($t_{1/2} \geq 4$ h) contained transcripts involved in housekeeping functions, while the mRNAs with short half-lives ($t_{1/2} < 4$ h) included regulatory mRNAs. Finally, we would like to emphasize that BRIC-seq also enables the characterization of RNA-binding proteins (RBPs), which may regulate the stability of their target transcripts. By combination with RNA-seq and RIP-seq/CLIP-seq, BRIC-seq will allow us to identify unknown RNA degradation pathways regulated by specific RBPs, such as UPF1 [17]. Although we have previously described the protocols for BRIC-seq [15, 18], we

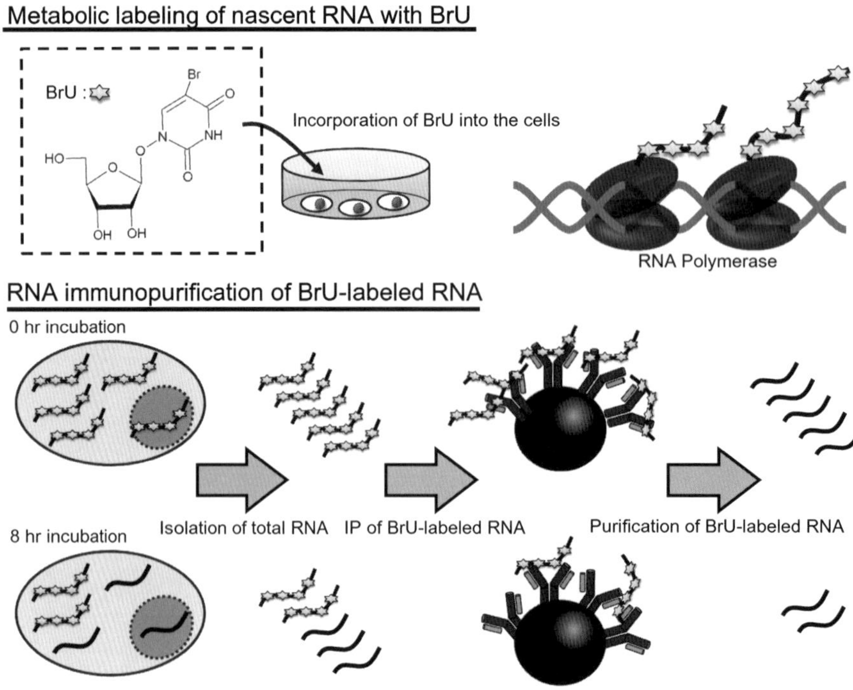

Fig. 1 Overview of the BRIC-seq protocol

herein present a modified analysis methodology that can determine the half-life of an individual transcript with high precision.

2 Materials

Prepare all solutions using RNase-free water (*see* **Note 1**). Unless mentioned otherwise, these solutions should be stored at 4 °C.

2.1 Metabolic Labeling

1. Dulbecco's modified Eagle's medium (DMEM)
2. Fetal bovine serum (FBS).
3. Antibiotics.
4. 100 mM 5′-bromouridine (BrU): Dissolve 323.1 mg of BrU in 10 ml of distilled water and sterilize through a 0.22-μm pore filter. Aliquot and store at −20 °C in the dark.
5. TRIzol Reagent (Life Technologies) (*see* **Note 2**).
6. Agilent RNA 6000 Nano Kit (Agilent Technologies).

2.2 Spike-In RNA

1. RiboMAX Large Scale RNA Production System SP6 and T7 (Promega). Store at −20 °C.
2. 1 mM 5′-bromouridine 5′-triphosphate sodium salt (BrUTP) (Sigma): dissolve 5.6 mg of BrUTP in 10 m of RNase-free water. Aliquot and store at −20 °C in the dark.
3. p-GEM Luc vector (Promega). Store at −20 °C.
4. QIAGEN RNeasy MinElute Cleanup Kit. Stored at room temperature (*see* **Note 3**).
5. Agilent RNA 6000 Nano Kit (Agilent Technologies).

2.3 RNA Immunopurification

1. Pierce Protein G agarose (Thermo Scientific) (*see* **Note 4**)
2. 10% bovine serum albumin (BSA). Dissolve 1 g of BSA in 10 mL of RNase-free water. Aliquot and store at −20 °C.
3. 10% Triton X-100: Mix 10 mL of Triton X-100 and 90 mL of RNase-free water. Store at room temperature.
4. BSA–Triton–PBS solution: Mix 5 mL of 10× Phosphate buffered saline (PBS) pH 7.4, 500 μL of 10% Triton X-100, 500 μL of 10% BSA, and 44 mL of RNase-free water. Prepare beforehand and store on ice.
5. 50 mg/mL heparin: dissolve 500 mg of heparin in 10 mL of RNase-free water. Aliquot and store at −20 °C.
6. Anti-Bromodeoxyuridine monoclonal antibody (mAb) 2B1 (Medical & Biological Laboratories Co., Ltd.) (*see* **Note 5**)
7. 2× TE Buffer (pH 7): Mix 1 mL of 1 M Tris–HCl pH 7.0, 200 μL of 0.5 M EDTA pH 8.0, and 48.8 mL of RNase-free water. Store at room temperature.

8. RNasin plus RNase inhibitor (Promega) (*see* **Note 6**).
9. Buffer A: Mix 500 μL of 1 M Tris–HCl pH 7.4, 500 μL of 5 M NaCl, and 49.0 mL of RNase-free water. Store at room temperature.
10. TRIzol LS Reagent (Life Technologies) (*see* **Note 7**).
11. Glycogen (Roche) (*see* **Note 8**).
12. Agilent RNA 6000 Pico Kit (Agilent Technologies).
13. 3 M NaOAc. Store at room temperature.
14. 100% EtOH. Store at room temperature.
15. 80% EtOH. Store at room temperature.

2.4 Quantification of RNA

1. TruSeq RNA Sample Prep Kit v2 (Illumina), which includes: Elute, Prime, Fragment Mix; First Strand Master Mix; Second Strand Master Mix; Resuspension buffer; A-Tailing Mix; DNA ligate Mix; RNA Adapter Mix; Ligation Mix; PCR Master Mix; and PCR Primer Cocktail.

3 Methods

Wear gloves for all of the steps and avoid RNase contamination. Carry out all procedures at room temperature, unless otherwise specified. Here we describe an experiment with six time points (0, 1, 2, 4, 8, and 12 h. *See* **Note 9**).

3.1 Metabolic Labeling of Nascent RNA by BrU

1. Prepare six 6 cm dishes with a cell line of choice (*see* **Note 10**). In each dish, seed the cells at 4×10^4 cells in 4 mL medium containing DMEM with 10% FBS and 1% Antibiotics. Incubate overnight at 37 °C and 5% CO_2.
2. For pulse-labeling of newly transcribed RNA with BrU, add BrU to a final concentration of 150 μM. After gently mixing, incubate for 24 h at 37 °C and 5% CO_2.
3. Wash the dishes twice with the fresh medium without BrU, followed by addition of 4 mL fresh medium.
4. Subsequently, total RNAs are isolated sequentially at specific time points over time after the medium change.
5. Lyse the cells by removing the medium and by adding 1.0 mL of TRIzol or RNAiso Plus directly onto the cells. Suspend the cells by pipetting up and down several times, and transfer the cell suspension to a 2-mL microcentrifuge tube (*see* **Note 11**).
6. Isolate the total RNA, which includes BrU-labeled RNAs and nonlabeled RNAs, according to the manufacturer's instructions.

7. Assess the RNA quality and concentration using an Agilent RNA 6000 Nano Kit on the Agilent Bioanalyzer 2100 (Agilent Technologies). An RNA Integrity Number (RIN) above 9 is desirable for subsequent experiments.

3.2 Preparation of Spike-In RNA

To evaluate the efficiency of RNA immunopurification, BrU- and nonlabeled synthesized RNAs, such as firefly luciferase (Luc) RNA, Renilla luciferase (RL) RNA, or green fluorescent protein (GFP), are used as a spike-in control RNA. The BrU-labeled RNAs are synthesized by in vitro transcription.

1. Mix 4 μL of T7 transcription 5× Buffer, 6 μL of rNTPs (25 mM ATP, CTP, GTP, UTP), 2 μL of linearized p-GEM Luc vector (800 ng/μL), 6 μL of 1 mM BrUTP, and 2 μL of Enzyme Mix (T7). Incubate for 4 h at 37 °C. In the case of synthesizing nonlabeled spike-in RNAs, BrUTP should be omitted from the reaction.
2. Add 1 μL of RQ1 RNase-Free DNase, and incubate for 15 min.
3. The transcribed RNAs are purified using the QIAGEN RNeasy MinElute Cleanup Kit, according to the manufacturer's instructions.
4. Assess the RNA quality and concentration using an Agilent RNA 6000 Nano Kit on the Agilent Bioanalyzer 2100.

3.3 RNA Immunopurification of BrU-Labeled RNA

1. Vortex Protein G agarose to resuspend.
2. Add 20 μL (multiplied by the number of RNA samples) bed volume of Protein G agarose in a 1.5-mL microcentrifuge tube.
3. Add 300 μL of ice-cold BSA/Triton/PBS solution. Mix well by inverting tubes several times. Centrifuge at 2500 × *g* for 3 min at 4 °C to pellet the agarose, and remove and discard the supernatant.
4. Repeat **step 3** twice.
5. Add 100 μL of ice-cold BSA/Triton/PBS solution, 10 μL of 50 mg/ml heparin, and 8 μg (multiplied by the number of RNA samples) of Anti-Bromodeoxyuridine mAb 2B1. Mix well by rotation at 4 °C for 2 h.
6. Centrifuge at 2500 × *g* for 3 min at 4 °C to pellet the agarose, and remove and discard the supernatant.
7. Add 1 mL of ice-cold BSA/Triton/PBS solution. Mix well by inverting tubes several times. Centrifuge at 2500 × *g* for 3 min at 4 °C to pellet the agarose, and remove and discard the supernatant.
8. Repeat **step 7** twice, and store at 4 °C until use.

9. Prepare 16 μg of BrU-labeled total RNA for each time point and 1 ng of BrU- and nonlabeled spike-in RNAs by diluting to a final volume of 100 μL with 1× TE buffer (pH 7).
10. Incubate these mixtures (**step 9**) for 2 min at 80 °C. Transfer the tubes to iced water immediately.
11. Add 100 μL of ice-cold BSA–Triton–PBS solution, 100 U of RNasin plus RNase inhibitor, and 10 μL of 50 mg/mL heparin to the antibody-conjugated protein G agarose. Resuspend the agarose by inverting several times.
12. Add heat-denatured BrU-labeled total RNA and spike-in RNAs. Mix well by rotation at 4 °C for 2 h.
13. Centrifuge at 2500 × *g* for 3 min at 4 °C to pellet the agarose, and remove and discard the supernatant. Be careful not to disturb the agarose that contain BrU-labeled RNAs.
14. Add 1 mL of ice-cold BSA–Triton–PBS solution. Mix well by inverting tubes several times. Centrifuge at 2500 × *g* for 3 min at 4 °C to pellet the agarose, and remove and discard the supernatant.
15. Repeat **step 15** three times.
16. Add 90 μL of buffer A and resuspend the agarose.
17. Add 300 μL of TRIzol LS Reagent and isolate the BrU-labeled RNA according to the manufacturer's instructions. During the step of formation of the RNA precipitant, add 60 μg of glycogen.
18. Assess the RNA quality and concentration using an Agilent RNA 6000 Pico Kit on the Agilent Bioanalyzer 2100.
19. Prepare 25 ng of BrU-labeled RNA by diluting to a final volume of 20 μL with RNase-free water.
20. Add 1 μL of glycogen, 2 μL of 3 M NaOAc, 60 μL of 100% EtOH to each sample and mix by pipetting up and down.
21. Centrifuge at maximum speed for 30 min at 4 °C, and remove and discard the supernatant.
22. Add 800 μL of 80% EtOH. Mix well by inverting tubes several times. Centrifuge at maximum speed for 3 min at 4 °C, and remove and discard the supernatant.
23. Air-dry the pellets for 2 min.
24. Elute the RNAs by adding 20 μL of RNase-free water (*see* **Note 12**).
25. Quantify the immunopurification efficiency by RT-PCR.

3.4 Quantification of RNA by Next Generation Sequencing

A TruSeq RNA Sample Prep Kit is used generate a high-quality cDNA library that is fragmented and ligated to an adapter, according to the manufacturer's instructions. The cDNA library is constructed without poly(A) selection. The cDNA library for BRIC-seq is sequenced using Illumina GAII or HiSeq, according to the manufacturer's instruction.

3.5 Data Analysis of Next Generation Sequencing

To perform BRIC-seq computational analysis, UNIX operating system such as Mac or Linux OS and more than about 8 GB (preferably 16 GB) of memory are required. In this section, we describe the computational analysis for BRIC-seq from preprocessing the raw sequence data (fastq files) to calculating the half-life for each transcript. Moreover, we present a workflow for analyzing significant changes of RNA half-lives between two conditions, such as transcripts in normal cell and transcripts in cells knockdown with specific RBP.

1. Remove low-quality reads and trim low-quality nucleotides from each FASTQ file (e.g., XXX_0h.fastq, XXX_1h.fastq, XXX_2h.fastq, XXX_4h.fastq, XXX_8h.fastq, XXX_12h.fastq,). To perform the quality filtering step in series, use the following command.

```
fastq_quality_filter -Q33 -q 20 -p 80 -i /path/to/
XXX_0h.fastq | fastq_quality_trimmer -Q33 -t 20 -l
10 -o /path/to/XXX_0h_filtered.fastq
```

2. Check the read quality on each fastq file using FastQC.

```
fastqc -o ./fastqc_XXX_0h ./XXX_0h.fastq -f fastq
```

3. Align the filtered reads to the rRNA reference sequences (*see* **Note 13**) to remove rRNA-derived reads and select unaligned reads for the next step.

```
bowtie -un /path/to/XXX_0h_norrna.fastq /path/to/Ribo-
somal_RNA /path/to/ XXX_0h_filtered.fastq > /dev/null
```

4. Align the preprocessed reads to the reference genome and a gene model (*see* **Note 14**) for each sample using Bowtie[19] and Tophat [20].

```
tophat -G /path/to/gencode.v19.annotation.gtf -o /
path/to/tophat_out_XXX_0h /path/to/hg19 /path/to/
XXX_0h_norrna.fastq
```

5. Estimate the RNA expression level (RPKM; Reads Per Kilobase per Million mapped reads) for each gene at each time point using Cufflinks [21]. The rpkm matrix (named genes.fpkm_table) that we use later is generated.

```
cuffnorm -p 8 --compatible-hits-norm -o /path/to/cuff-
norm_out_XXX -G /path/to/ gencode.v19.annotation.gtf /
path/to/ tophat_out_XXX_0h/accepted_hits.bam /path/to/
tophat_out_XXX_1h/accepted_hits.bam /path/to/topha-
t_out_XXX_2h/accepted_hits.bam /path/to/topha-
t_out_XXX_4h/accepted_hits.bam
/path/to/tophat_out_XXX_8h/accepted_hits.bam
/path/to/tophat_out_XXX_12h/accepted_hits.bam
```

6. Estimate a fitting curve for the time points of each gene, calculate the RNA half-life for each gene and then compare the change of RNA half-lives between two conditions using custom R library named BridgeR (https://github.com/AkimitsuLab/BridgeR, *see* **Note 15**). Briefly, the relative expression data of each transcript at each time point (compared with 0 h) is normalized using the appropriate percentile (e.g., 95% or 97.5%, Figs. 2a, b and 3). To predict a RNA decay curve, the normalized expression data is fitted to first order exponential RNA decay model. RNA half-life and its variance are estimated from the fitting curve and time points (Fig. 2c). To determine the significant change of RNA half-lives, the distributions of RNA half-lives between two conditions are compared by using the test statistics (Welch's *t*-test, Fig. 2d). We prepared the sample scripts in github repository (https://github.com/AkimitsuLab/scripts_for_BRIC-seq_data_analaysis). In the following example, we showed the sample commands for calculating RNA half-life calculation and finding the significant change of RNA half-lives between two conditions. At first, the rpkm matrix data including only mRNA data or ncRNA data are extracted from the original rpkm matrix data (named genes.fpkm_table file).

```
For mRNA,
filename_1="cuffnorm_out_XXX_Control"
filename_2="cuffnorm_out_XXX_Knockdown"
gene_list="/path/to/gencode.v19.annotation_symbol_ty-
pe_list.txt"
python make_map_from_gtf.py ${gtfFile} ${gene_list}
cuffnorm_data_1="./${filename_1}/genes.fpkm_table"
cuffnorm_data_2="./${filename_2}/genes.fpkm_table"
gene_type="mRNA"
result_file_1="BridgeR_input_file_mRNA_${filename_1}.
txt"
result_file_2="BridgeR_input_file_mRNA_${filename_2}.
txt"
python BridgeR_prep.py ${gene_list} ${cuffnorm_data_1}
${gene_type} ${result_file_1}
python BridgeR_prep.py ${gene_list} ${cuffnorm_data_2}
${gene_type} ${result_file_2}
```

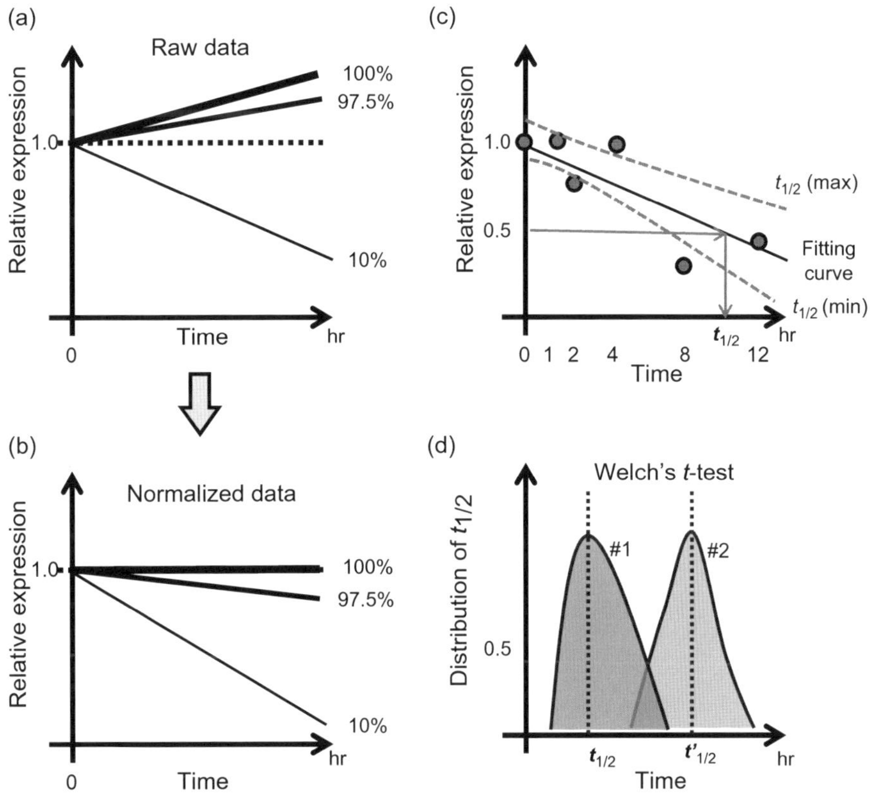

Fig. 2 Workflow of the analysis step. (**a**) The percentile of a score calculated from the RPKM value for all transcript data in BRIC-seq is plotted against time. Values of 100%, 97.5%, and 10% are shown. (**b**) Normalized percentile of a score versus time. The percentile of each score shown in (**a**) is normalized using an appropriate percentile (e.g., 95% or 97.5%). (**c**) The relative expression (transformed to $\log_{10}$) of each transcript at each time point is plotted against time (h). The plot is a curve fitted with a first-order exponential RNA decay model. The broken lines represent the upper ($t_{1/2}$ (max)) and lower ($t_{1/2}$ (min)) standard deviation around the mean. (**d**) Distributions of $t_{1/2}$ in different conditions (#1 and #2). A p-value is calculated between the two distributions based on Welch's t-test

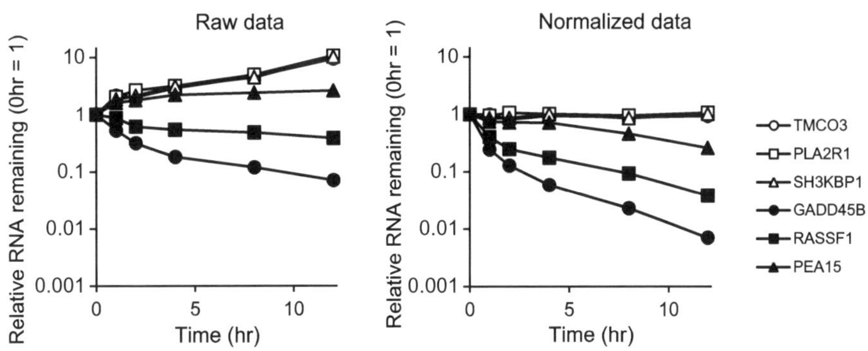

Fig. 3 Representative RNA decay plot. The relative RNA amounts of representative transcripts were plotted against time before (left) and after (right) the normalization

```
For ncRNA,
filename_1="cuffnorm_out_XXX_Control"
filename_2="cuffnorm_out_XXX_Knockdown"
gene_list="/path/to/gencode.v19.annotation_symbol_ty-
pe_list.txt"
cuffnorm_data_1="./${filename_1}/genes.fpkm_table"
cuffnorm_data_2="./${filename_2}/genes.fpkm_table"
gene_type="lncRNA"
result_file_1="BridgeR_input_file_lncRNA_${file-
name_1}.txt"
result_file_2="BridgeR_input_file_lncRNA_${file-
name_2}.txt"
python BridgeR_prep.py ${gene_list} ${cuffnorm_data_1}
${gene_type} ${result_file_1}
python BridgeR_prep.py ${gene_list} ${cuffnorm_data_2}
${gene_type} ${result_file_2}
```

Next, to calculate RNA half-life and find the significant change of RNA half-lives between two conditions, use the following commands.

```
For mRNA,
result_file_1="BridgeR_input_file_mRNA_${filename_1}.txt"
result_file_2="BridgeR_input_file_mRNA_${filename_2}.txt"
Rscript
BridgeR_analysis_mRNA_for_DEG.R ./${result_file_1} ./
${result_file_2}
For ncRNA,
result_file_1="BridgeR_input_file_lncRNA_${filename_1}.txt"
result_file_2="BridgeR_input_file_lncRNA_${filename_2}.txt"
Rscript BridgeR_analysis_lncRNA_for_DEG.R ./${result_file_1}
./${result_file_2} ../BRIC-seq_BridgeR_mRNA/BridgeR_3_Norma-
lizaion_factor_Control_mRNA.txt
```

Finally, "BridgeR_6_HalfLife_Pvalue_estimation.txt" file is created in current directory. This file includes RNA half-lives in two conditions and the p-value of the test statistics (Welch's t-test).

4 Notes

1. As the protocol involves handling RNA, make sure that you are working in an RNase-free environment. The operator should wear RNase-free gloves, and all the reagents, tubes, and tips used must be RNase free.

2. Equivalent reagents can be used, such as RNAiso plus (TAKARA).
3. Equivalent reagents can be used, such as MicroSpin G-25 (GE).
4. Protein G agarose is critical for purifying BrU-labeled RNA at high quality and quantity. In our experience, Pierce Protein G agarose (Thermo Scientific) works best.
5. Selection of antibody is critical in purification of BrU-labeled RNA. To our knowledge, the anti-Bromodeoxyuridine mAb 2B1 supplied from Medical & Biological Laboratories Co., Ltd. showed the best recovery rate of BrU-labeled RNA with low nonspecific binding compared with other antibodies.
6. Equivalent reagents can be used, such as RNaseOUT recombinant Ribonuclease Inhibitor (Thermo Scientific).
7. Equivalent reagents can be used, such as ISOGEN LS (NIPPON GENE).
8. Equivalent reagents can be used, such as Ethachinmate (NIPPON GENE).
9. The number of time points determines the range and accuracy of half-life determination by BRIC-seq. Here, we observed six time points (0, 1, 2, 4, 8, and 12 h), which are reliable to measure half-lives between 1 and 24 h.
10. Not all types of cell lines incorporate BrU. As far as we have examined, transcripts of HeLa, A549, DLD1, TIG3, and HEK293 cell lines are efficiently labeled with BrU. Suitable labeling conditions for BrU are those where the recovery of BrU-labeled RNAs is ten times higher than that of nonlabeled RNAs.
11. Safe stopping Point—Protocol may be stopped here, if users decide to not proceed to the next step. If users are stopping the protocol, store the sample at −80 °C.
12. We highly recommend confirming half-life of specific transcripts by reverse transcription-quantitative real-time polymerase chain reaction (RT-qPCR). Briefly, the RNA is reverse transcribed into cDNA using the PrimeScript RT Master Mix (Perfect Real Time, TAKARA). The cDNA is then amplified by SYBR Premix Ex Taq II (Perfect Real Time, TAKARA). Spike-in RNAs (Luc RNAs) or housekeeping gene such as GAPDH or PGK1 mRNA.
13. 18S and 28S rRNA fasta files are prepared from NCBI website and then merge two fasta files using cat command.

```
cat 18S_rRNA.fasta 28S_rRNA.fasta > rRNA.fasta
```

To build an index file named "Ribosomal_RNA" of rRNAs to align sequence reads to the rRNA using bowtie mapping software, use the following command.

```
bowtie-build rRNA.fasta Ribosomal_RNA
```

14. A prebuilt index file of a reference genome is prepared from the Bowtie website (http://bowtie-bio.sourceforge.net/index.shtml) A gene model annotation such as RefSeq or GENCODE is also prepared from iGenomes (http://support.illumina.com/sequencing/sequencing_software/igenome.html) or Gencode website (https://www.gencodegenes.org/).
15. We prepared an R script for BridgeR library installation (named BridgeR_installation.R script) in github repository (https://github.com/AkimitsuLab/scripts_for_BRIC-seq_data_analaysis). For the first time, use the following command to install BridgeR library in your computer.

```
Rscript ./BridgeR_installation.R
```

We also prepared shell script and python and R scripts for RNA half-life calculation. In github repository (https://github.com/AkimitsuLab/scripts_for_BRIC-seq_data_analaysis), there are two folders named "half-life_calculation" and "half-life_calculation_for_DEG_analysis" to calculate RNA half-life and find the significant change of RNA half-lives between two conditions, respectively.

Acknowledgments

Toshimichi Yamada and Naoto Imamachi contributed equally to this work.

References

1. Ing NH (2005) Steroid hormones regulate gene expression posttranscriptionally by altering the stabilities of messenger RNAs. Biol Reprod 72:1290–1296
2. Staton J et al (2000) Hormonal regulation of mRNA stability and RNA-protein interactions in the pituitary. J Mol Endocrinol 25:17–34
3. Ge Q et al (2003) RNA interference of influenza virus production by directly targeting mRNA for degradation and indirectly inhibiting all viral RNA transcription. Proc Natl Acad Sci U S A 100:2718–2723
4. Voinnet O (2005) Induction and suppression of RNA silencing: insights from viral infections. Nat Rev Genet 6:206–220
5. Maity A et al (1995) Evidence for post-transcriptional regulation of cyclin B1 mRNA in the cell cycle and following irradiation in HeLa cells. EMBO J 14:603–609
6. Wang W et al (2000) HuR regulates cyclin A and cyclin B1 mRNA stability during cell proliferation. EMBO J 19:2340–2350
7. Raghavan A et al (2002) Genome-wide analysis of mRNA decay in resting and activated

primary human T lymphocytes. Nucleic Acids Res 30:5529–5538

8. Selinger DW et al (2003) Global RNA half-life analysis in Escherichia coli reveals positional patterns of transcript degradation. Genome Res 13:216–223
9. Grigull J et al (2004) Genome-wide analysis of mRNA stability using transcription inhibitors and microarrays reveals posttranscriptional control of ribosome biogenesis factors. Mol Cell Biol 24:5534–5547
10. Narsai R et al (2007) Genome-wide analysis of mRNA decay rates and their determinants in Arabidopsis thaliana. Plant Cell 19:3418–3436
11. Wang Y et al (2002) Precision and functional specificity in mRNA decay. Proc Natl Acad Sci U S A 99:5860–5865
12. Rabani M et al (2011) Metabolic labeling of RNA uncovers principles of RNA production and degradation dynamics in mammalian cells. Nat Biotechnol 29:436–442
13. Schwanhausser B et al (2011) Global quantification of mammalian gene expression control. Nature 473:337–342
14. Windhager L et al (2012) Ultrashort and progressive 4sU-tagging reveals key characteristics of RNA processing at nucleotide resolution. Genome Res 22:2031–2042
15. Imamachi N et al (2014) BRIC-seq: a genome-wide approach for determining RNA stability in mammalian cells. Methods 67:55–63
16. Tani H et al (2012) Genome-wide determination of RNA stability reveals hundreds of short-lived noncoding transcripts in mammals. Genome Res 22:947–956
17. Tani H et al (2012) Identification of hundreds of novel UPF1 target transcripts by direct determination of whole transcriptome stability. RNA Biol 9:1370–1379
18. Tani H et al (2015) Genome-wide analysis of long noncoding RNA turnover. In: Nakagawa S, Hirose T (eds) Nuclear bodies and noncoding RNAs: methods and protocols. Springer, New York, pp 305–320
19. Langmead B et al (2009) Ultrafast and memory-efficient alignment of short DNA sequences to the human genome. Genome Biol 10:R25
20. Trapnell C et al (2009) TopHat: discovering splice junctions with RNA-Seq. Bioinformatics 25:1105–1111
21. Trapnell C et al (2010) Transcript assembly and quantification by RNA-Seq reveals unannotated transcripts and isoform switching during cell differentiation. Nat Biotechnol 28:511–515

Chapter 2

Determining mRNA Decay Rates Using RNA Approach to Equilibrium Sequencing (RATE-Seq)

Farah Abdul-Rahman and David Gresham

Abstract

RATE-seq is a *4-thiouracil* (4-tU)-based method that enables the in vivo measurement of transcriptome-wide RNA degradation rates. 4-tU is an analog of uracil that is rapidly incorporated into newly synthesized RNA and facilitates the conjugation of a biotinylated molecule containing a reactive thiol group. The biotinylated RNA can then be fractionated from the unlabeled RNA with streptavidin magnetic beads. By adding 4-tU to a culture of cells growing in steady-state conditions, fractionating the labeled population of RNA at multiple time points following 4-tU addition, and quantifying the abundance of newly transcribed RNAs using RNAseq, it is possible to estimate the degradation rates of all transcripts in a single experiment. The analysis of the RATE-seq data entails normalization of RNAseq libraries to thiolated RNA spike-ins and nonlinear model fitting to estimate the degradation rate constant for each RNA species.

Key words 4-thiouracil, RNA degradation, RNA turnover, Metabolic labeling, RNA stability

1 Introduction

Measuring RNA degradation rates poses two main challenges: (1) minimizing perturbation of cellular physiology and (2) efficiently assaying thousands of transcripts to assess transcriptome-wide dynamics. Historically, most methods for measuring RNA decay rates involved halting transcription, through chemical or genetic inhibition of RNA polymerase, and tracking the decrease in RNA levels over time [1–3]. However, such methods are typically deleterious to cells and thereby are potentially confounded by stress-induced changes in RNA metabolism and gene expression. More recently, the metabolic label 4-thiouracil (4-tU), a uracil analogue that is rapidly incorporated into newly synthesized RNA, has been used in two different ways to study transcriptome-wide degradation rates: using either a pulse-chase or a pulse-in experimental design [4–6]. In a pulse-chase experiment, 4-tU is added to a growing cell population for several generations. The 4-tU label is then chased with unlabeled uracil and the rate of

Shireen R. Lamandé (ed.), *mRNA Decay: Methods and Protocols*, Methods in Molecular Biology, vol. 1720,
https://doi.org/10.1007/978-1-4939-7540-2_2, © Springer Science+Business Media, LLC 2018

decrease in labeled RNA is used to estimate degradation kinetics. In a pulse-in experiment, the rate with which RNA is labeled and reaches an equilibrium value is used to estimate transcript degradation rates. Pulse-chase methods are potentially susceptible to effects of nucleotide recycling, which may result in the reincorporation of the 4-tU label from degraded transcripts into newly synthesized RNAs. An aberrantly "slow chase" can result, leading to an underestimation of mRNA decay rates. By contrast, in pulse-in methods this problem is circumvented by quantifying the incorporation kinetics of 4-tU.

*R*NA *A*pproach to *E*quilibrium *S*equencing (RATE-seq) combines a pulse-in design with time course sampling and RNA sequencing (RNAseq) to estimate transcriptome-wide degradation rates. Briefly, multiple RNA samples are collected at defined time points after 4-tU is pulsed into a culture of cells growing in steady-state conditions. Following RNA purification, a cocktail of synthetic thiolated spike-ins is added to each time point sample in a fixed ratio of spike-in to RNA quantity. After conjugation to a biotin-containing reactive molecule (either biotin-HPDP or biotin-MTS [7]), the labeled RNA is fractionated using streptavidin magnetic beads, cDNA libraries are generated, and the entire pool of cDNA is processed using standard RNAseq protocols [4].

The underlying principle of RATE-seq is that the abundance of any mRNA species can be modeled by the mathematical relationship: $d[\text{mRNA}]/dt = k - \alpha \cdot [\text{mRNA}]$, where [mRNA] is the concentration of a particular mRNA, k is the rate of transcript synthesis, α is the exponential decay constant, and t is time. On the basis of this model, the abundance of any mRNA at time (t) is given by the relationship $[\text{mRNA}](t) = [\text{mRNA}]_{ss} \cdot (1 - e^{\alpha \cdot t})$, where $[\text{mRNA}]_{ss}$ is the concentration of a given mRNA in steady-state conditions [8, 9]. When the labeled fraction is considered independent of the unlabeled fraction, the same relationship holds for labeled RNA: $[\text{mRNA}_{\text{labeled}}](t) = [\text{mRNA}_{\text{labeled}}]_{ss} \cdot (1 - e^{\alpha \cdot t})$. A feature of this model is that the time an RNA takes to reach its equilibrium value is a function only of α. In general, an RNA that reaches its equilibrium value quickly has a fast degradation rate (i.e., a larger value of α) whereas an RNA that reaches its equilibrium value slowly has a slow degradation rate (i.e., a smaller value of α). RATE-seq aims to determine the value of the exponential degradation rate constant, α, by fitting this model to each mRNA using spike-in-normalized RNAseq counts of the labeled mRNAs over time. Here, we describe the experimental procedures for generating RATE-seq data and provide guidelines for the analysis of data using normalization of RNAseq data to exogenous thiolated spike-ins and nonlinear model fitting. Our protocol focuses on 4-tU labeling and fractionation. A protocol for RNAseq library preparation has not been provided as any standard RNAseq protocol can be employed. RATE-seq has been developed in yeast cells, but is amenable to any cell type that transports and incorporates 4-tU.

2 Materials

Prepare all solutions using ultrapure water (prepared by purifying deionized water, to attain a sensitivity of 18 MΩ-cmat 25 °C) and analytical grade reagents. Prepare and store all reagents at room temperature (unless indicated otherwise). Follow all waste disposal regulations when disposing of waste materials.

2.1 Sampling

1. 4-tU labeling reagent: Dissolve 4-tU in DMSO so that the final concentration is 500 μM in the cell culture.
2. 25mm, 0.45 micron circular filters.
3. Liquid nitrogen.
4. Filtration device (vacuum flask, vacuum source, glass microanalysis filter holder assembly).
5. 2 mL eppendorf tubes.

2.2 RNA Extraction

1. Lysis buffer: 0.01 M EDTA, 0.5% SDS, 0.01 M Tris pH 7.5, RNase-free H_2O.
2. Acid phenol.
3. Heavy phase lock gel (PLG) tubes.
4. Chloroform.
5. Sodium acetate: 3 M sodium acetate. pH with glacial acetic acid to 5.2.
6. 95% ethanol.
7. 70% ethanol.
8. Qubit RNA HS kit and Qubit fluorometer.

2.3 Making In Vitro Spike-ins

1. *Escherichia coli* strains with plasmids DGP 104, 105, 106 (*see* **Note 1**).
2. LB media: 10 g Bacto tryptone, 5 g yeast extract, and 10 g NaCl in 1 L of water.
3. EcoRI restriction enzyme.
4. 10× EcoRI compatible buffer.
5. Qiagen Miniprep Kit.
   (a) 100 mM DTT.
6. Promega Riboprobe in vitro Transcription System.
   (a) rNTP.
   (b) SP6 RNA Polymerase.
   (c) Transcription optimized 5× buffer.
7. Qiagen PCR Cleanup Kit.
8. 10 mM 4-thiouridine.
9. Qubit DNA HS.

2.4 Fractionation of Labeled RNA

1. 1 M Tris–HCl, pH 7.4.
2. 0.5 M EDTA.
3. 1 M HEPES.
4. Biotinylation reagent: Dissolve 10 mg of Biotin MTS in 100 mL of Dimethylformamide.
5. 75% ethanol.
6. Chloroform–Isoamyl alcohol (24:1).
7. Isopropoanol.
8. 5 M NaCl.
9. Glycogen.
10. RiboZero Kit.
11. Magnetic stand.
12. Streptavidin magnetic beads.
13. 5% Beta-mercaptoethanol.
14. 1.5 mL eppendorf tubes.
15. Streptavidin bead buffer: 1 M NaCl, 10 mM EDTA, 100 mM Tris–HCl, pH 7.4.
16. Reagents for RNAseq protocol of choice.

3 Methods

Carry out all procedures at room temperature unless otherwise specified. Use RNAse-free H_2O unless otherwise specified.

3.1 Sampling

1. Grow liquid culture in defined media containing a total of 500 μM uracil until exponential growth.
2. Assemble vacuum flask with filter support and turn on vacuum source. Test by running ultra pure water onto the filter to ensure that the liquid is being sucked out.
3. Add 4-tU solution to the growing culture to a final concentration of 500 μM. Sample the first time point by pipetting 10 mL of culture into the filter apparatus. Remove the circular filter with tweezers once the liquid has drained. Place the filter into a 2 mL eppendorf tube and immediately place in liquid nitrogen. Samples can be stored in −80 C for subsequent processing.
4. Rinse filter apparatus with ultrapure water in between samplings and continue sampling as in **step 3** for approximately 2 h. Samples should be acquired frequently at the beginning of the time course and less frequently toward the end to capture approach to equilibrium kinetics. An example time course would be sampling at 3, 5, 7, 11, 25, 35, 50, and at 160 min.

3.2 RNA Extraction

1. Add 750 μL of lysis buffer and 750 μL of acid phenol to the frozen samples and vortex until foamy and opaque. Incubate for 1 h at 65 °C and vortex every 20 min.
2. Discard the filter and put the samples on ice for 10 min.
3. Spin samples for 5 min at 5000 × *g*.
4. With a pipette, transfer the top aqueous layer to a prespun PLG tube.
5. Add 750 μL chloroform and invert to mix (*see* **Note 2**) and spin at 5000 × *g* for 5 min.
6. Pour aqueous layer into a new 2 mL eppendorf (*see* **Note 3**).
7. Add 75 μL of 3 M sodium acetate and 1.2 mL 100% ethanol, then vortex samples. Incubate samples at −80 °C for 30 min (*see* **Note 4**).
8. Spin samples at 5000 × *g* for 10 min at 4 °C and decant supernatant.
9. Wash the surface of the pellet with 400 μL of 70% ethanol and spin samples for 2 min at 5000 × *g* and discard supernatant. Repeat this step then proceed to **step 10**.
10. Air-dry samples inverted on the bench for 30 min.
11. Dissolve pellet in 100 μL water by pipetting up and down (*see* **Note 5**).

3.3 Making Synthetic Spike-In RNAs

1. Grow up *E. coli* strains DGP 104, 105, 106 in LB media overnight.
2. Follow the Qiagen miniprep protocol to extract plasmids. Quantify using Qubit DNA HS.
3. Linearize plasmids by making a reaction of 5 μL of 10× EcoRI compatible buffer, 5 μL EcoRI, 3 μg of the extracted plasmid to a total volume of 50 μL. Incubate for 2 h at 37 °C (*see* **Note 6**).
4. Clean up reactions using Qiagen PCR cleanup kit and resuspend in 30 μL of H_2O. Quantify using Qubit DNA HS. Around 100–200 ng/mL is required to proceed.
5. Add 6.25 μL of linearized plasmid DNA, 4 μL transcription optimized 5× buffer, 2 μL DTT, 0.75 μL Recombinant RNasin, 4 μL rNTP, 2 μL 4sUTP, and 1 μL of SP6 RNA Polymerase. Incubate for 1 h at 30 °C.
6. Add 1 μL RQ1 RNase-Free DNase and incubate for 15 min at 37 °C.
7. Add 19 μL acid phenol. Vortex vigorously and centrifuge at 12,000 × *g* for 2 min.
8. Transfer top aqueous layer to a new eppendorf tube and add 1.8 μL of 3 M Na Acetate and 45 μL of 100% ethanol (*see* **Note 3**). Incubate for 30 min in −80 °C (*see* **Note 4**).

9. Centrifuge samples at full speed at 4 °C for 25 min and discard supernatant.
10. Wash with 1 mL 70% ethanol then centrifuge for 5 min at full speed. Discard supernatant and repeat this step then proceed to **step 11**.
11. Resuspend pellet in 10 μL of H_2O by pipetting up and down.

3.4 Fractionation of Labeled RNA (See Note 7)

1. Dissolve RNA in 20 μL of water to a concentration of 500 ng/μL.
2. Add 147 μL H_2O, 2.5 μL 1 M HEPES, 0.5 μL 0.5 M EDTA, and 50 μL RNA sample. Add the 50 μL of biotin last and vortex briefly. Incubate samples in the dark for 2 h.
3. Add 250 μL chloroform–isoamyl alcohol (24:1) and pipet up and down until mixed (*see* **Note 8**).
4. Transfer into prespun PLG tube and centrifuge for 5 min.
5. Pour the aqueous phase of the PLG into a new 1.5 mL eppendorf (*see* **Note 9**).
6. Precipitate RNA by adding 1/10 volume of 5 M NaCl and 1.1 volume of isopropanol and incubate at room temperature for 10 min.
7. Spin for 20 min at full speed at 4 °C and discard the supernatant (*see* **Note 10**).
8. Add 400 μL of 75% ethanol and spin for 5 min at full speed at 4 °C and remove supernatant. Resuspend in H_2O to a concentration of 500 ng/μL.
9. Warm 100 μL of bead buffer per sample to 65 °C.
10. Add 200 μL beads to new eppendorf tubes and place in magnetic rack. Leave for 1.5 min until beads sediment and then discard the supernatant.
11. Remove tubes from magnetic rack and wash the beads with 200 μL bead buffer by pipetting up and down. Sediment beads in magnetic rack again for 1.5 min and discard supernatant.
12. Remove tubes from magnetic rack and resuspend in 100 μL of bead buffer. Add 100 μL of RNA sample (*see* **Note 11**) and incubate for 20 min on the bench.
13. Place samples in magnetic stand for 2 min then discard the supernatant (*see* **Note 12**).
14. Remove tubes from magnetic rack and resuspend in 100 μL of bead buffer and incubate for 5 min. Sediment beads in magnetic rack again for 2 min and discard supernatant.
15. Remove tubes from magnetic rack and resuspend in 100 μL bead buffer and incubate for 1 min. Sediment beads in magnetic rack again for 2 min and discard supernatant.

16. Remove tubes from magnetic rack and resuspend in 100 μL of the 65 °C prewarmed bead buffer and leave for 1 min. Sediment beads in magnetic rack again for 2 min and discard supernatant.
17. Remove tubes from magnetic rack and resuspend in 100 μL bead buffer and leave for 1 min. Sediment beads in magnetic rack again for 2 min and discard supernatant.
18. Remove tubes from magnetic rack and resuspend beads in 20 μL of 5% β-mercaptoethanol and incubate for 10 min at room temperature. Sediment beads in magnetic rack again for 2 min and transfer supernatant to a new eppendorf (*see* **Note 13**).
19. Repeat β-mercaptoethanol incubation at 65 °C with the beads from **step 18**, then pool that with the supernatant from **step 18**.
20. Precipitate RNA with 1/10 volume 5 M NaCl, 4 μg glycogen, and 1.1 volume Isopropanol.
21. Incubate for 10 min at room temperature and spin at maximum speed for 25 min at 4 °C.
22. Discard supernatant and wash with 75% ethanol then spin at maximum speed for 10 min.
23. Resuspend pellet in 10 μL water. Sediment beads in magnetic rack again for 2 min and transfer supernatant to a new eppendorf tube.
24. Quantify RNA concentration using Qubit RNA HS (*see* **Note 14**).
25. Use the RiboZero Kit as per instructions to remove rRNA (*see* **Note 15**).
26. Generate RNAseq libraries using a standard strand-specific RNAseq protocol.

3.5 Computational Analysis

1. Align reads to reference transcriptome using a splice-aware aligner (e.g., HiSat2). The reference genome, and annotation file (if used), must contain the sequences of the synthetic spike-in RNAs. Generate a BAM file containing mapped reads.
2. Count the number of reads mapping to each transcript using a program such as htseq-count or Rsubread and generate a matrix containing gene name and number of RNAseq counts at each time point. The annotation file (gtf or gff) used for this step must contain the relevant information for the spike-ins.
3. Test whether spike-ins decrease in relative abundance compared to native RNA counts with time (*see* Fig. 1).
4. Normalize RNAseq counts within each timepoint to spike-in abundance by determining the total proportion (p) of reads mapping to all spike-ins in the library i (p_i) and dividing the RNAseq counts for each RNA j in library i by p_i (RNA_{ji}/p_i).

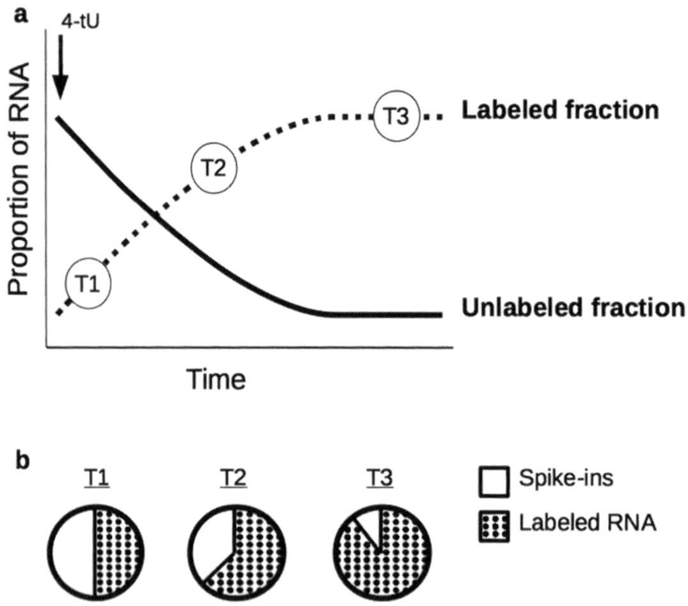

Fig. 1 Principle of RNA approach to equilibrium sequencing. (**a**) Fractionating the labeled RNA from a sample time course collected after the addition of 4-tU yields increasing amounts of RNA until equilibrium is reached. T1, T2, and T3 are time points representing the beginning, middle, and end of a typical RATE-seq time course. (**b**) Pie charts representing fraction of RNA-seq reads belonging to spike-ins compared to RNA. While the absolute quantity of spike-ins is the same for all time points, its fraction shrinks as the labeled RNA fraction increases. Normalizing RNA-seq reads to a common pool of synthetic spike-ins accounts for technical variation that is attributable to variation in RNA-seq library sizes between time points

5. Fit a nonlinear model to the normalized counts to estimate approach to equilibrium kinetics using a statistical programming language, such as R. Using regression, estimate the parameters $[\mathrm{mRNA}_{\mathrm{labeled}}]_{\mathrm{ss}}$ and α using the formula $[\mathrm{mRNA}_{\mathrm{labeled}}](t) = [\mathrm{mRNA}_{\mathrm{labeled}}]_{\mathrm{ss}} \cdot (1 - e^{\alpha \cdot t})$, where $[\mathrm{mRNA}_{\mathrm{labeled}}]_{\mathrm{ss}}$ is the amount of labeled RNA at steady state, α is the combined sum of growth rate and RNA degradation rate, and t is time. Extract the degradation rate constant for each mRNA.
6. To assess model fits, plot normalized data and model fits (*see* Fig. 2).

4 Notes

1. Synthetic spike-ins were generated from *Bacillus subtilis* and *Caenorhabditis elegans* sequences cloned into a polyA vector to generate polyA-tailed RNAs using in vitro transcription. For sequence information see [4].
2. Vortexing will ruin PLG tubes.

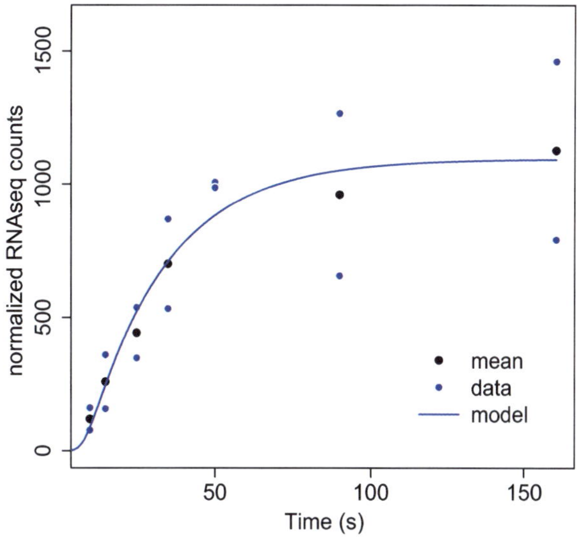

Fig. 2 Approach to equilibrium profile of normalized counts. A nonlinear model is fit to the data using the equation: $[\text{mRNA}_{\text{labeled}}](t) = [\text{mRNA}_{\text{labeled}}]_{\text{ss}} \cdot (1 - e^{\alpha \cdot t})$ to estimate the values of $[\text{mRNA}_{\text{labeled}}]_{\text{ss}}$ and α. Model fitting is ideally performed using a weighted regression that aims to account for unequal variances between time points or a maximum likelihood method that assumes negative binomial distributions of counts at each time point

3. If your sample has low amounts of RNA, use glycogen as a carrier to help precipitate the RNA.
4. Samples can also be precipitated overnight by incubating at −20 °C.
5. Begin by resuspending in a small amount of water then dilute further if the concentration is too high.
6. Run linearized and unlinearized fragments on a 1% gel to check that the reaction has proceeded as expected.
7. A dotblot should be performed as in ref. 4 to monitor the dynamics of label incorporation into bulk RNA. Once it is established that 4-tU is being incorporated into RNA as expected in your system, it is not necessary to repeat a dotblot each time RATE-seq is used.
8. Alternate RNA precipitation methods can be used here such as an ethanol precipitation or an Ampure bead cleanup.
9. You can try to pipet out the remnants of the aqueous phase unless there is white particulate inside.
10. The pellet does not stick well to the tube and could be discarded with the supernatant if the sample is not handled carefully.

11. If adding more than 50 μg, make sure the volume of bead buffer to sample is at least 1:1.
12. If interested in the unlabeled RNA, save the unbound fraction for downstream processing and sequencing.
13. The liquid should be pink/light purple color. Avoid accidentally transfering beads into the new tube.
14. An increase in RNA concentration with time should be observed.
15. This kit is expensive. Scaling down this protocol by a half has been successful.

References

1. Wiesner RJ, Zak R (1991) Quantitative approaches for studying gene expression. Am J Phys 260(4 Pt 1):L179–L188
2. Wang Y, Liu CL, Storey JD, Tibshirani RJ, Herschlag D, Brown PO (2002) Precision and functional specificity in mRNA decay. Proc Natl Acad Sci U S A 99:5860–5865. https://doi.org/10.1073/pnas.092538799
3. Grigull J, Mnaimneh S, Pootoolal J, Robinson MD, Hughes TR (2004) Genome-wide analysis of mRNA stability using transcription inhibitors and microarrays reveals posttranscriptional control of ribosome biogenesis factors. Mol Cell Biol 24:5534–5547. https://doi.org/10.1128/MCB.24.12.5534-5547.2004
4. Munchel SE, Shultzaberger RK, Takizawa N, Weis K (2014) Dynamic profiling of mRNA turnover reveals gene-specific and system-wide regulation of mRNA decay. Mol Biol Cell 22:2787–2795. https://doi.org/10.1091/mbc.E11-01-0028
5. Miller C, Schwalb B, Maier K, Schulz D, Dümcke S, Zacher B, Mayer A, Sydow J, Marcinowski L, Dölken L, Martin DE, Tresch A, Cramer P (2011) Dynamic transcriptome analysis measures rates of mRNA synthesis and decay in yeast. Mol Syst Biol 7:58. https://doi.org/10.1038/msb.2010.112
6. Neymotin B, Athanasiadou R, Gresham D (2014) Determination of in vivo RNA kinetics using RATE-seq. RNA 20:1645–1652. https://doi.org/10.1261/rna.045104.114
7. Duffy EE, Rutenberg-Schoenberg M, Stark CD, Kitchen RR, Gerstein MB, Simon MD (2015) Tracking distinct RNA populations using efficient and reversible covalent chemistry. Molecular Cell 59(5)858–866
8. Greenberg JR (1972) High stability of messenger RNA in growing cultured cells. Nature 240:102–104
9. Kim CH, Warner JR (1983) Messenger RNA for ribosomal proteins in yeast. J Mol Biol 165:79–89

Chapter 3

Metabolic Labeling of Newly Synthesized RNA with 4sU to in Parallel Assess RNA Transcription and Decay

Wei Sun and Wei Chen

Abstract

The development of genome-wide RNA profiling technologies greatly facilitates the global analysis of gene expression. However, such technologies alone could not distinguish the contribution to cellular RNA abundance by transcription versus decay. To overcome such limitation, metabolic labeling of newly synthesized RNA with 4-thiouridine (4sU) combined with genome-wide RNA profiling was used to in parallel measure RNA transcription and decay. Here, we describe the detailed protocol for using metabolic labeling with 4sU to separate newly synthesized RNA from the preexisting RNA in mammalian cells.

Key words 4-Thiouridine(4sU), Metabolic labeling, Newly synthesized RNA, RNA transcription, RNA decay, Global analysis

1 Introduction

Gene expression is a multistep process. Due to the technical convenience, in a genome-wide manner, one of the most popular means to analyze gene expression is to measure cellular RNA abundances [1, 2]. However, RNA abundances are determined by the balance between their synthesis and decay. Measuring RNA abundances alone could not differentiate the contributions from the two processes [3, 4]. Importantly, the control of RNA transcription and RNA decay results from different classes of regulatory mechanisms. Therefore, to understand the molecular mechanisms underlying gene expression regulation, it is necessary to develop suitable methods to dissect the contributions of RNA transcription and RNA decay.

Recently, one approach based on metabolic labeling of newly synthesized RNA in living cells with 4-thiouridine (4sU) was developed to enable the direct assessment of RNA transcription [3–7]. 4sU is a naturally existing uridine derivative. After exposing to cells, 4sU can be rapidly taken up and directly incorporated into RNA transcripts by the cellular transcription machinery with

Shireen R. Lamandé (ed.), *mRNA Decay: Methods and Protocols*, Methods in Molecular Biology, vol. 1720,
https://doi.org/10.1007/978-1-4939-7540-2_3, © Springer Science+Business Media, LLC 2018

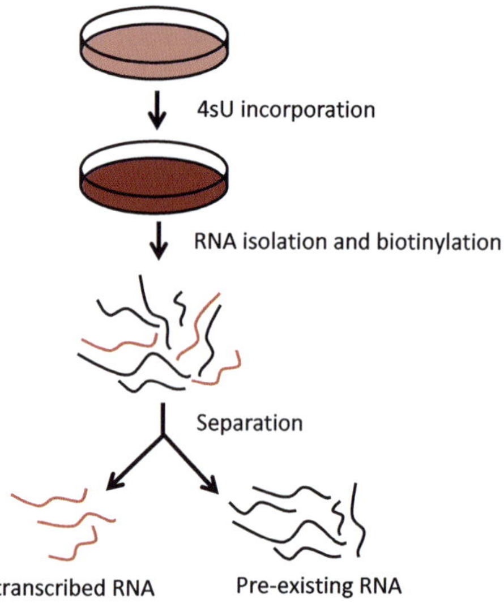

Fig. 1 Experimental design of metabolic labeling with 4sU. 4sU is incorporated into cells by adding into culture medium for the required time. The 4sU labeled total RNA is isolated and thiol-specifically biotinylated. Then the 4sU-biotinylated total RNA is separated into 4sU-biotin-labeled, newly transcribed RNA and unlabeled, preexisting RNA by streptavidin-mediated separation

minimal influence on cell status and gene expression [8] . Following total RNA isolation, the 4sU-containing newly synthesized RNAs can be specifically labeled with biotin via thiol-specific biotinylation. Thereafter, by streptavidin mediated separation, cellular RNA can be separated into 4sU-labeled "newly transcribed" RNA and unlabeled "preexisting" RNA (Fig. 1). Combining this 4sU labeling with genome-wide RNA profiling approaches, such as gene expression microarray and RNA deep-sequencing, genome-wide measurement of RNA transcription and decay can be achieved in parallel.

Here, we describe the detailed protocol for metabolic labeling and isolation of the newly transcribed RNA in cultured mammalian cells. After separating the newly transcribed and the preexisting RNA, this method can allow direct global analysis of the dynamics of RNA synthesis, processing, and degradation for eukaryotic cells in steady state [3, 4] or under certain stimulation [3, 5–7].

2 Materials

2.1 Reagents

Prepare all solutions using ultrapure water (prepared by purifying deionized water, to attain a sensitivity of 18 MΩ-cm at 25 °C) and analytical grade reagents. Prepare and store all reagents at room temperature (RT) unless indicated otherwise. Strictly follow all waste disposal regulations when disposing waste materials.

1. 4sU solution: Use 4sU powder (Sigma-Aldrich, T4509) to prepare the solution as 50 mM stock in H_2O. Store the solution at −20 °C in aliquots of 100 μl (*see* **Note 1**).
2. TRIzol (Invitrogen) (*see* **Note 2**).
3. Chloroform (*see* **Note 3**).
4. 25 G needles.
5. Phase Lock Gel tube, 1.5 ml (5 PRIME).
6. Isopropanol.
7. 5 M NaCl.
8. 80% ethanol. Store at −20 °C.
9. 100% ethanol. Store at −20 °C.
10. RNase-free water.
11. Biotin-HPDP solution: Prepare 1 mg/ml stock solution by dissolving 50 mg of biotin-HPDP (Pierce) in 50 ml of dimethylformamide (DMF). Store at 4 °C in aliquots of 1 ml.
12. 1 M Tris–HCl (pH 7.4)
13. 0.5 M EDTA.
14. Phenol–chloroform–isoamyl alcohol (25:24:1, v/v) (*see* **Notes 2** and **3**).
15. Positively charged nylon membranes (Amersham™ Hybond™-N+ Membrane).
16. Dot blot blocking buffer: PBS, pH 7.5, 10% SDS, 1 mM EDTA.
17. Streptavidin-horseradish peroxidase (ThermoFisher). Store at −20 °C.
18. Dot blot washing buffer 1: PBS, pH 7.5, 10% SDS.
19. Dot blot washing buffer 2: PBS, pH 7.5, 1% SDS.
20. Dot blot washing buffer 3: PBS, pH 7.5, 0.1% SDS.
21. ECL reagent (GE Healthcare).
22. μMacs Streptavidin Kit (Miltenyi). Store the buffer and beads at 4 °C.
23. 100 mM dithiothreitol (DTT). Freshly prepare before use.
24. Separation washing buffer: 100 mM Tris–HCl, pH 7.4, 10 mM EDTA, 1 M NaCl, 0.1% Tween 20.
25. 3 M NaOAc (pH 5.2).
26. Glycogen, RNA grade (ThermoFisher).

2.2 Equipment

Here, only the equipment that a standard molecular biology laboratory does not possess is listed.

1. UV cross-linker: Stratalinker 2400 UV Crosslinker, 365 nm (Stratagene).
2. Magnetic stand (Miltenyi).
3. Agilent Bioanalyzer 2100 (Agilent).

3 Methods

Carry out all procedures at RT unless otherwise specified.

3.1 Metabolic Labeling with 4sU

Before starting, plan the experimental procedure in details, including when to add 4sU to cell culture and when to collect the 4sU labeled cells (*see* **Note 4**).

During 4sU handling, avoid exposing it (including 4sU labeling cells) to bright light.

1. Just before use, thaw and add required amount of 4sU solution into a sterile Falcon tube (*see* **Note 5**).
2. Transfer the required amount of cell culture medium (5 ml per 10 cm dish) from the cell culture dishes into the 4sU containing Falcon tube and mix thoroughly.
3. Remove and discard the remaining medium in the cell culture dishes, and add the 4sU containing medium back to the cell culture dishes.
4. At the end of labeling time, remove the cell culture medium, and add 4 °C-cold TRIzol reagent into cell culture dishes (5 ml per 10 cm dish). Collect the cell lysate and transfer to 15 ml or 50 ml falcon tube. Homogenize by first vortexing and then passing lysate through a 25 G needle 8–10 times (TRIzol is extremely hazardous, *see* **Note 2**).
5. Incubate at RT for 5 min. At this point the samples can also be frozen at −80 °C.

3.2 Total RNA Isolation

This procedure of total RNA isolation was modified based on standard TRIzol RNA extraction protocol (Invitrogen).

1. For 1 ml of TRIzol lysate, add 200 μl of chloroform to each tube (*see* **Note 3**).
2. Shake vigorously for 15 s. Incubate at RT for 2–3 min. Transfer to a prespun Phase Lock Gel tube and spin 15 min at ≥12,000 × *g* at 4 °C to separate phases (*see* **Note 6**).
3. Transfer upper aqueous phase to fresh 1.5 ml RNase-free tube (contain RNA). Add equal volume of isopropanol and 1/10 volume of 5 M NaCl, invert to mix.
4. Incubate 10 min at RT or overnight at −20 °C to precipitate RNA. Collect the RNA pellet by spinning at ≥12,000 × *g* for 10 min at 4 °C.

5. Carefully remove supernatant and DO NOT disturb the RNA pellet at the bottom of the tube (*see* **Note 7**).
6. Wash pellet with 1 ml of 80% ethanol without disturbing the pellet (*see* **Note 8**).
7. Spin at 7500 × *g* for 5 min at 4 °C.
8. Remove supernatant, pulse-spin to remove residual ethanol and air-dry pellets 2–5 min at RT (*see* **Note 9**).
9. Dissolve pellets in 20–40 μl of RNase-free water (*see* **Note 10**). At this point the samples can be frozen and stored at −80 °C.
10. Determine RNA concentration and purity by taking OD 260 and 280 values.

3.3 Thio-Specific Biotinylation of 4sU Containing Newly Transcribed RNA

1. Prepare a 1 mg/ml solution of biotin-HPDP in DMF.
2. Assemble the 4sU-labeled total RNA biotinylation reaction in a 1.5-ml microfuge tube. Example reaction for 50 μg of 4sU labeled total RNA:

RNase-free water	344 μl
1 M Tris–HCl (pH 7.4)	5 μl
0.5 M EDTA	1 μl
1 μg/μl 4sU labeled total RNA	50 μl
1 mg/ml biotin-HPDP	100 μl (must be added last)
Total volume	500 μl

3. Incubate the biotinylation reaction at RT for 2–3 h in dark.
4. Purify the biotinylated RNA from the excess of biotin-HPDP using Phase-Lock-Gel tubes. Add the same amount of phenol-–chloroform–isoamyl alcohol (*see* **Notes 2** and **3**), mix it vigorously in the tube and centrifuge at full speed for 5 min at 4 °C.
5. Transfer the upper aqueous phase which containing RNA. Precipitate the labeled RNA by addition of 1/10 reaction volume of 5 M NaCl and 1.1 volume of isopropanol. Incubate at RT for 10 min.
6. Spin in a microcentrifuge at 20,000 × *g* for 20 min at 4 °C to collect the RNA pellet at the bottom. Discard the supernatant.
7. Add 500 μl of 80% ethanol to wash the pellet and respin in a microcentrifuge at 20,000 × *g* for 5 min at 4 °C.
8. Dissolve RNA in 100 μl (for every 50 μg of total RNA) of RNase-free water. At this point the samples can be frozen and stored at −80 °C.
9. Determine RNA concentration and purity by taking OD 260 and 280 values (*see* **Note 11**).

3.4 Dot Blot Checking 4sU Incorporation

4sU incorporation can be checked by dot blotting biotinylated RNA and probing with streptavidin-horseradish peroxidase. This dot blot experiment is optional.

1. Drop 2 μl of 4sU-biotinalyted RNA (around 1 μg) onto a positively charged nylon membrane (*see* **Note 12**).
2. UV crosslink the membrane twice with 1200 μJ using UV cross-linker machine.
3. Block the membrane by incubating in dot blot blocking buffer for 20 min.
4. Probe the membrane by incubating with 1:10,000 dilution of 1 mg/ml streptavidin-horseradish peroxidase in dot blot blocking buffer for 15 min.
5. Wash the membrane twice with dot blot washing buffer 1 for 10 min each time.
6. Wash the membrane twice with dot blot washing buffer 2 for 10 min each time.
7. Wash the membrane twice with dot blot washing buffer 3 for 10 min each time.
8. Visualize the signal of biotin-bound HRP by ECL detection reagent according to manufacturer's protocol. Typical dot blot result is shown in Fig. 2.

3.5 Separation of 4sU-Biotinylated RNA from Unlabeled RNA

1. In each low-binding tube, use 100 μl of 4sU-biotinylated RNA (450–500 ng/μl) for each μMacs column, but increase the volume to 200 μl by adding additional 100 μl of RNase-free water.
2. Denature RNA samples at 65 °C for 10 min followed by rapid cooling on ice for 5 min.
3. For each tube, add 200 μl of μMacs Streptavidin MicroBeads, and mix and rotate for 15 min at RT (*see* **Note 13**).

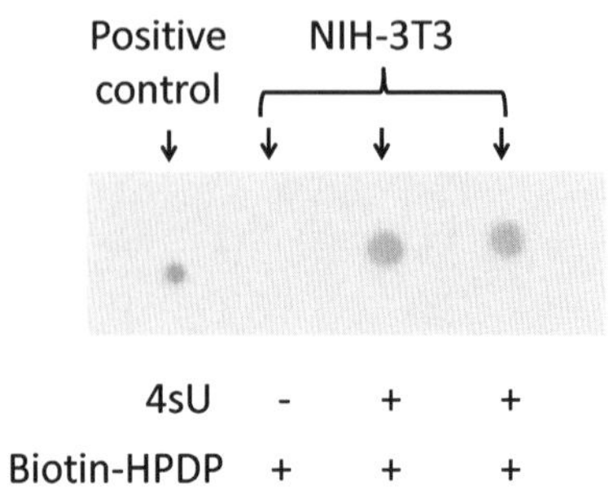

Fig. 2 Dot blot to check 4sU incorporation. Two-hundred micromolar 4sU was added into the culture medium of NIH-3T3 cells for 2 h. One plate of NIH-3T3 cells was cultured without adding 4sU, serving as negative control. Positive control was 5′-biotin labeled DNA oligo, as mentioned in **Note 12**

4. Take the nuclear-acid equilibration buffer out of 4 °C fridge and warm back to RT. Prepare the fresh 100 mM DTT in RNase-free water. Almost at the end of the incubation for beads/RNA, put the μColumn in the magnetic field of the μMacs magnetic stand. Prepare the column by rinsing with 2 × 100 μl of the nuclear-acid equilibration buffer (from μMacs Streptavidin Kit).
5. After washing column, put a new collecting tube under the column (*see* **Note 14**).
6. Shortly spin down the Beads-RNA sample, and gently resuspend the beads by pipetting up and down three times and load 200 μl samples onto the top of the column matrix: The beads (bound with 4sU-biotinalyted RNA) are magnetically retained within the column while the preexisting RNA are flowed out and collected in the collecting tubes from **step 5**.
7. Spin briefly to collect the preexisting RNA and load back onto the column two additional times (three loadings in total).
8. Wash the column with 200 μl of 55 °C separation washing buffer and collect the wash buffer into the collecting tube with preexisting RNA from **steps 5** to 7 (*see* **Note 15**).
9. Change the collection tube and wash the column for additional two times with more than 300 μl of 55 °C separation washing buffer. Discard the flow-through washing buffer.
10. After discard the washing buffer, put a new collecting tube under the column (*see* **Note 16**).
11. Elute the biotin-4sU labeled RNA from the column by adding 100 μl of freshly prepared 100 mM DTT, and collect into the collecting tubes from **step 10**. Then perform a similar second- and third-time elution and collections with 5 min incubation in between. The column should be retained in the magnetic field during the whole procedure.
12. Recover the preexisting RNA and newly transcribed RNA by ethanol precipitation. Add 3 μl of glycogen for each newly transcribed RNA tube (*see* **Note 17**), mix by tapping the tube, and then add 1/10 volume of 3 M NaOAc (pH 5.2), mix by tapping the tube, and finally add 3× volume of cold 100% ethanol, mix by inverting the tubes for several times (*see* **Note 18**).
13. Incubate the tubes for more than 1 h at −80 °C or overnight at −20 °C.
14. Spin in a microcentrifuge at 20,000 × *g* for 20 min at 4 °C to collect the RNA pellet at the bottom. Discard the supernatant.

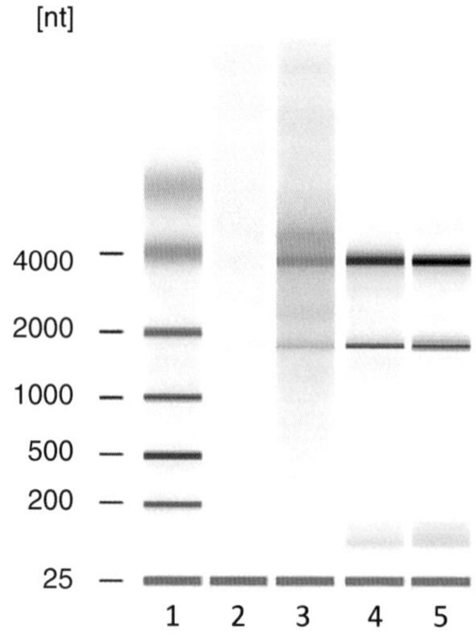

Fig. 3 Electrophoretic analysis for the newly transcribed, preexisting, and total RNA after separation. Analysis from Agilent Bioanalyzer 2100 system using RNA 6000 Nano kit. 200 μM 4sU was added into the culture medium of NIH-3T3 cells for 2 h. One plate of NIH-3T3 cells was cultured without adding 4sU, serving as negative control. Lane 1: ladder; Lane 2: 4sU-labeled newly transcribed RNA from negative control (cells without 4sU adding); Lane 3–5: 4sU-labeled newly transcribed RNA (lane 3), unlabeled preexisting RNA (lane 4), and total RNA (lane 5) from 4sU labeled cells

15. Dissolve the RNA pellet with appropriate amounts of RNase-free water (*see* **Note 19**). At this point the samples can be frozen and stored at −80 °C.
16. Determine RNA concentration and purity by taking OD 260 and 280 values.
17. Check RNA integrity by electrophoretic analysis to exclude RNA degradation (*see* **Note 20**). Typical result is shown in Fig. 3.

4 Notes

1. Do not refreeze aliquots of 4sU solution when used. Discard any leftovers from thawed aliquots.
2. For phenol and phenol containing reagents (such as TRIzol). WARNING: CORROSIVE AND HAZARDOUS TO HEALTH! Ensure immediate access to Phenol antidote (PEG-Methanol, store at 4 °C).
3. For chloroform and chloroform containing reagents. WARNING: HAZARDOUS TO HEALTH!

4. Although the minimum 4sU labeling time and interval cab be as short as 5 min [3, 6], we recommend to allow at least 15 min for the labeling, and at least 15 min intervals between different labeling times. For standard newly transcribed RNA/preexisting RNA separation, 1–2 h of 4sU labeling time would be recommended. Handle maximum of four dishes of cells at a given labeling time point. Handle cells as quickly as possible to minimize the influences on temperature and CO_2 levels, and guarantee the accuracy of 4sU labeling times.
5. Normally, we label cells with 4sU in medium at a final concentration of 100–200 μM. Higher concentrations of 4sU may influence the cell growth [9].
6. Prespin the Phase Lock Gel tube at 12,000 × *g* for 2 min, RT. And read through manufacturer's protocol before using Phase Lock Gel tube.
7. RNA pellet will be very white.
8. Pellet will often float free. Also RNA pellets can be stored in 80% ethanol at −80 °C for up to 1 year safely.
9. Do not overdry the pellet. Extensively drying RNA pellets makes them hard to resuspend.
10. If RNA does not resuspend completely, you may not have a good yield and more RNase-free water can be added.
11. The concentration of the biotin-RNA should be 450–500 ng/μl.
12. Optionally, one biotin positive control could be included. Such positive control could be 1 μl of 10 ng/μl 25 nt random DNA oligo with 5′-biotin.
13. The ratio between the volumes of beads and 4sU-biotinylated RNA should be 1:1; approximately 100 μl of μBeads for around 25 μg of labeled RNA and in total 50 μg RNA/200 μl Beads for each column. Binding is completed in seconds, thus prolonged incubation is not necessary.
14. This collecting tube is for collecting flow-through preexisting RNA.
15. The total volume of preexisting RNA is around 600 μl per tube after collecting. Split each sample into 2 × 1.5 ml low-binding tubes for the RNA precipitation in following steps.
16. This collecting tube is for collecting flow-through newly transcribed RNA.
17. The glycogen is not necessary for precipitation of preexisting RNA.
18. The volumes of split preexisting RNA and newly transcribed RNA are around 300 μl in each tube.

19. Normally, add 30 μl of water for each sample of newly transcribed RNA and 100 μl of water for each sample of split preexisting RNA. Let the pellet dissolve for more than 3 min and then pipetting up and down and combine same samples into one tube.
20. Typically, we use the Agilent Bioanalyzer 2100 system to for analysis.

References

1. Wang Z, Gerstein M, Snyder M (2009) RNA-Seq: a revolutionary tool for transcriptomics. Nat Rev Genet 10:57–63
2. Brown PO, Botstein D (1999) Exploring the new world of the genome with DNA microarrays. Nat Genet 21:33–37
3. Dölken L, Ruzsics Z, Rädle B et al (2008) High-resolution gene expression profiling for simultaneous kinetic parameter analysis of RNA synthesis and decay high-resolution gene expression profiling for simultaneous kinetic parameter analysis of RNA synthesis and decay. RNA 14:1–13
4. Schwanhäusser B, Busse D, Li N et al (2011) Global quantification of mammalian gene expression control. Nature 473:337–342
5. Rabani M, Levin JZ, Fan L et al (2011) Metabolic labeling of RNA uncovers principles of RNA production and degradation dynamics in mammalian cells. Nat Biotechnol 29:436–442
6. Windhager L, Bonfert T, Burger K et al (2012) Ultrashort and progressive 4sU-tagging reveals key characteristics of RNA processing at nucleotide resolution. Genome Res 22:2031–2042
7. Rabani M, Raychowdhury R, Jovanovic M et al (2014) High-resolution sequencing and modeling identifies distinct dynamic RNA regulatory strategies. Cell 159:1698–1710
8. Hafner M, Landthaler M, Burger L et al (2010) Transcriptome-wide identification of RNA-binding protein and microRNA target sites by PAR-CLIP. Cell 141:129–141
9. Burger K, Mühl B, Kellner M et al (2013) 4-thiouridine inhibits rRNA synthesis and causes a nucleolar stress response. RNA Biol 10:1623–1630

Chapter 4

Measuring mRNA Decay in Budding Yeast Using Single Molecule FISH

Tatjana Trcek, Samir Rahman, and Daniel Zenklusen

Abstract

Cellular mRNA levels are determined by the rates of mRNA synthesis and mRNA decay. Typically, mRNA degradation kinetics are measured on a population of cells that are either chemically treated or genetically engineered to inhibit transcription. However, these manipulations can affect the mRNA decay process itself by inhibiting regulatory mechanisms that govern mRNA degradation, especially if they occur on short time-scales. Recently, single molecule fluorescent in situ hybridization (smFISH) approaches have been implemented to quantify mRNA decay rates in single, unperturbed cells. Here, we provide a step-by-step protocol that allows quantification of mRNA decay in single *Saccharomyces cerevisiae* using smFISH. Our approach relies on fluorescent labeling of single cytoplasmic mRNAs and nascent mRNAs found at active sites of transcription, coupled with mathematical modeling to derive mRNA half-lives. Commercially available, single-stranded smFISH DNA oligonucleotides (smFISH probes) are used to fluorescently label mRNAs followed by the quantification of cellular and nascent mRNAs using freely available spot detection algorithms. Our method enables quantification of mRNA decay of any mRNA in single, unperturbed yeast cells and can be implemented to quantify mRNA turnover in a variety of cell types as well as tissues.

Key words smFISH, mRNA decay, Single cells, Single molecule, Budding yeast, Yeast, *S. cerevisiae*, Fluorescent in situ hybridization, mRNA turnover, mRNA half-life, mRNA decay rate

1 Introduction

Eukaryotic gene expression is a complex enzymatic process, during which an mRNA is transcribed in the nucleus and exported into the cytoplasm where it is translated and finally degraded [1]. In *Saccharomyces cerevisiae*, mRNA half-lives ($t_{1/2}$) vary considerably; highly unstable mRNAs show $t_{1/2} \sim 3$ min while highly stable ones show a $t_{1/2}$ of >90 min [2–4]. Long-lived mRNAs can undergo many more rounds of translation compared to short-lived transcripts before being degraded. Therefore, the control of mRNA turnover constitutes an additional layer of regulation to optimize protein levels in cells [5].

Shireen R. Lamandé (ed.), *mRNA Decay: Methods and Protocols*, Methods in Molecular Biology, vol. 1720,
https://doi.org/10.1007/978-1-4939-7540-2_4,

To measure mRNA half-lives, mRNA decay needs to be uncoupled from mRNA synthesis such that a time-dependent change in cellular mRNA levels will only reflect the decaying mRNA population. Afterward, mRNA decay is described by the first-order decay kinetics, where the change in the mRNA concentration at a given time (dN/dt) depends on the concentration of the decaying mRNA (N). This relationship is summarized by the following equation [3, 6]:

$$\frac{dN}{dt} = -k_d N, \tag{1}$$

where k_d is the decay constant of an mRNA. Solving this equation gives an integrated rate equation:

$$\ln\left(\frac{N_t}{N_0}\right) = -k_d t, \tag{2}$$

where N_t is the amount of mRNA at time t and N_0 is the amount of mRNA at time zero. Thus, in the absence of transcription, a change in the mRNA abundance is a direct reflection of mRNA decay expressed by the decay rate constant k_d [3, 6]. k_d translates into mRNA half-life using Eq. 3:

$$t_{1/2} = \frac{0.693}{k} \tag{3}$$

where mRNA half-life is the time it takes for a given mRNA concentration to be reduced by 50%. It indicates a kinetic stability of an mRNA species; the longer the mRNA half-life, the more stable the mRNA.

Three different categories of approaches have been used to quantify mRNA decay rates in budding yeast. The first approach includes methods that use transcriptional inhibitors, temperature sensitive RNA polymerase II (RNAP II) mutants or inducible promoters to uncouple transcription from mRNA decay and therefore rely on chemically treated or genetically perturbed cells [6, 7]. The second category includes methods that use modified nucleotides to label mRNAs in vivo in minimally genetically perturbed cells without the use of transcriptional inhibitors. These two categories are often coupled with high throughput sequencing for a genome-wide characterization of mRNA turnover [6–9]. The last category includes the single molecule fluorescent in situ hybridization (smFISH) approach, which allows quantification of mRNA stability in single, minimally genetically or chemically perturbed cells. smFISH is not a genome-wide approach, however, it enables quantification of changes in mRNA stability without transcriptional inhibition and without normalization of mRNA signal to extract the kinetic information of a decaying mRNA. By doing so, smFISH achieves high temporal resolution to record changes in mRNA stability that occur on short timescales and are cell cycle-dependent [10]. In this protocol, we describe in detail how to use smFISH to determine mRNA half-lives in single yeast cells.

In budding yeast, chemical inhibition of transcription using thiolutin and 1,10-phenantroline is the most widely used approach to study mRNA turnover [6, 7]. However, studies showed that inhibition of transcription using these drugs is not complete; thiolutin inhibits expression approximately 95% efficiently while 1,10-phenantroline inhibits transcription approximately 90% efficiently [6]. Furthermore, different strains show variable sensitivity to transcription inhibitors, and variable effects on different functional groups of genes have also been observed [11]. Transcription can also be inhibited using thermally labile alleles of RNAP II, such as temperature sensitive ***rbp1–1*** mutation. mRNA synthesis is inhibited by shifting a yeast culture from permissive to nonpermissive temperature, which inactivates RNAP II activity approximately 90% efficiently [4, 6]. After transcription is turned off, a decay curve is obtained through a time-dependent sampling of cells. Total cellular RNA is isolated and relative levels of decaying mRNA species are quantified using Northern blot, qRT-PCR, microarray analysis or high throughput sequencing. Relative mRNA levels are obtained by normalizing the mRNA levels of a gene of interest against a reference mRNA, usually a long lived mRNA. As described above, decay measurements are obtained by mathematical fitting of decay curves using Eqs. 2 and 3 [4, 6, 12–14].

In general, inhibition of transcription using drugs or temperature sensitive RNAP II alleles broadly affects cellular physiology. For example, thiolutin likely inhibits cellular processes other than mRNA synthesis, and at high concentrations can affect the decay process itself [6, 9, 11]. Furthermore, temperature-sensitive RNAP II alleles cannot be coupled with other conditional mutants while the heat shock itself can have secondary effects on the mRNAs stability [6]. Both approaches can reduce the cellular pool of labile factors that are involved in mRNA decay making the interpretation of results obtained by these methods challenging. To circumvent these problems, transcription of a target mRNA can be temporally controlled using inducible promoters, such as galactose- or tetracycline-inducible promoters [7]. These promoters provide a switch between transcriptional induction and inhibition without broadly disrupting cellular physiology. However, construction of special strains is required and some yeast strains could be tetracycline-sensitive. Additionally, a change of carbon source to induce galactose-dependent expression will differentially affect mRNA stability of transcripts associated with a particular metabolic pathway. Importantly, mRNA stability of some genes has been shown to be entirely regulated by their promoters rather than specific *cis* regulatory sequences embedded in the mRNA [10, 15, 16], indicating that mRNA turnover should be quantified on mRNAs expressed from endogenous promoters.

To circumvent the use of drugs and temperature sensitive mutants, metabolic labeling approaches have been introduced to globally measure decay rates. Typically, radioactive nucleosides and more recently 4-thiouridine (4sU) are used to label cellular mRNA. During a short pulse, cells take up and incorporate modified nucleosides into newly synthesized transcripts. Following a short labeling pulse, the medium containing modified nucleosides is removed and replaced with a fresh media containing an excess of unlabeled nucleosides to "chase" the intracellular labeled mRNA pool. Thus, without transcriptional inhibition, mRNA decay of multiple mRNAs can be monitored in parallel [3, 6–9]. Large amounts of radioactive material are needed to adequately label cellular mRNA, thus 4sU is becoming more widely used for mRNA labeling. However, to allow efficient 4sU uptake, yeast cells have to be genetically engineered to express a nucleoside transporter that enables 4sU uptake [8, 9]. The particular advantage of this method is that when combined with high throughput sequencing, mature mRNAs as well as nascent mRNA are labeled concurrently to asses global mRNA synthesis and decay in minimally chemically and genetically perturbed cells [8, 9].

In this protocol we discuss how, by counting absolute numbers of cellular mRNAs and nascent mRNAs at the site of transcription, the mRNA decay rate of a genetically unmodified gene can be calculated using mathematical modeling. We and others have shown that normalization of mRNA decay data with a reference mRNA to measure mRNA turnover combined with transcriptional inhibition prolong mRNA half-lives by as much as twofold [9, 10]. By avoiding normalization and transcriptional inhibition, smFISH enables quantification of mRNA decay in chemically and genetically unperturbed cells. Unlike traditional techniques, which rely on a population of cells to quantify mRNA levels, smFISH enables quantification of mRNA turnover in single cells. Thus, variation in gene expression occurring in a fraction of cells, otherwise obscured by ensemble measurements, can be readily detected [10]. This method also revealed that for some cell cycle-regulated mRNAs such as *SWI5* and *CLB2*, their cell cycle-dependent mRNA decay is coordinated with their transcription and that the specificity for mRNA turnover is entirely encoded by their promoter regions and not by specific sequences encoded in the mRNA [10]. In this protocol we describe how the smFISH approach can be coupled with mathematical modeling to quantify the mRNA stability of a non-cell-cycle regulated, housekeeping gene such as *MDN1*. After hybridization with smFISH probes, absolute number of cellular mRNAs and nascent mRNAs associated with the active transcription site are quantified using freely available spot detection algorithms. A steady state established between mRNA synthesis and mRNA decay is assumed to determine mRNA half-lives. While the mRNA staining protocol described here is applicable to fluorescently label mRNAs in budding yeast, the spot detection and quantification of mRNA turnover can be employed in other organisms, such as mammalian cell lines and tissues [17]. As quantitative smFISH is becoming increasingly used in the RNA field, we believe that the protocol described here will become a valuable resource for the study of mRNA turnover in intact biological samples.

2 Materials

All solutions are prepared using ultra-pure, double distilled water and analytical grade reagents. Formamide is toxic and should be handled in a fume hood using protective gloves and discarded according to the environmental health and safety instructions. All solutions are prepared at room temperature (RT). Maintain DNase- and RNase-free conditions during hybridization and use DNase- and RNase-free reagents to prepare the solutions.

1. *32% paraformaldehyde* (Electron Microscopy Sciences).
2. *70% ethanol.* To make 100 ml of 70% ethanol, combine 70 ml of 100% ethanol and 30 ml of dH_2O.
3. *20× saline-sodium citrate buffer* (20× SSC; Roche).
4. *Deionized formamide* (Amresco).
5. *1× Buffer B.* 1 l of 1× Buffer B contains 1.2 M sorbitol and 100 mM potassium phosphate buffer pH 7.5, dissolved in ddH_2O. Store at 4 °C.
6. *1.4× Buffer B.* 1 l of 1.4× Buffer B contains 1.7 M sorbitol and 140 mM potassium phosphate buffer pH 7.5, dissolved in ddH_2O. Store at 4 °C.
7. *Spheroplast Buffer.* 2.5 ml of spheroplast buffer contains 1.8 ml of 1.4× buffer B, 5 μl of β-mercaptoethanol, 250 μl of 200 mM Ribonucleoside Vanadyl Complex (VRC) (New England Biolabs) and 0.39 ml ddH_2O. Prepare fresh.
8. *Poly-L-lysine coated microscope coverslips.* Thoroughly wash microscope coverslips by boiling them in 10 mM HCl on the stirring hotplate for 10 min. Discard the solution and wash repeatedly with ddH2O. Air-dry on Whatman paper. Coat each coverslip with 200 μl of 1× (0.01%) poly-L-lysine (Sigma), incubate for 10 min, aspirate and air-dry. Wash 2× with ddH_2O and let air-dry. Store at room temperature (RT).
9. *smFISH probe mix.* Commercially available Stellaris® FISH Probes (single-stranded DNA oligonucleotides) against a target gene are custom designed using the Stellaris® RNA FISH Probe Designer (Biosearch Technologies, Inc.; www.

biosearchtech.com/stellarisdesigner). In this study, we designed 48 probes, each 20-nucleotides-long, against the 5′ end of the *MDN1* mRNA. Their 5′–3′ sequences were as follows: cagagggaaaagcagaattg, ttgttgctaaagtggaaggg, cggctatgtagtatagttcc, gaaaacatgaccagtgatgg, caacgtgccactatctc-taa, ggcattgcaacgggaaatat, gaatccttttgtgtggatgg, ggttgaat-gaagcgtgcaaa, ccaacaatgaatcgcgtgat, tccagaatcgccctcaaaat, gttcgaggaagctgtaatca, gcacatctcttagcttcgtt, gaacactgggtagttta-gag, cctgaggctcaataatgaag, attgacgcgaccttagtact, tttgtcgtgga-tagtgtgga, ggacaaaggttaatgggtag, cgaagagaggaaaccgtttt, ccgcttttccaatgagcatt, gagtcatggcaacccatata, ggcatcagtttgtt-caccta, caccagaggtataagtacca, acaccagctctccattcaaa, tacc-catctcccttctttga, cgcgcttttctaaaagcgat, tcctctggatggaatggtta, catttgcagcctttacagtc, cgctaggttcctctaattca, ggttggtcaaaatgg-gaaac, cccccttgttcaatgaaatg, ctaacctttcgcacagctta, caggtttgtt-gatgccattg, cgtagacagaagactggatt, tgaattctccaatagcgcca, ctcgccgattgcttgtataa, ccttgaggaagcaatgtcta, ggcacatgttgggt-caaaaa, gcgaacaatgtgctgttcat, ccgactagtaaaacaggttc, tgaacgactg-tagttttccc, ccaagaagatcaccagtttc, gcatcttgtgaaacttctcg, gtacgcttcgttccaaagtt, cagcccatttgtcaagtaac, cctcaaacttcttcact-gag, ccctcgacaaaattgaagac, gccctgatagtctttaccaa, ttcatcgagcaa-tagccact. The probes are synthesized with a free functional amine (mdC(TEG-Amino)) at the 3′ or 5′ end, which can be conjugated with an amine reactive fluorescent dye. To detect single cytoplasmic *MDN1* mRNAs and nascent *MDN1* chains at the site of active transcription, smFISH probes hybridizing to different positions along the 5′ end of the target mRNA were used (*see* **Note 1**) (Fig. 1a) [18]. This approach allows quantification of the absolute number of cellular mRNAs as well as the absolute number of nascent transcripts associated with an actively transcribed gene.

In this method, *MDN1* probes are coupled to a cyanine 3 (Cy3) dye, however other dyes such as CAL Fluor590, Quasar 670 and Cy5 can also be used (*see* **Note 2**). To conjugate smFISH probes with Cy3 fluorophores (Sigma-Aldrich), unlabeled smFISH probes are purchased and coupled to cy3 post synthesis. Probes are delivered individually in ddH_2O (here, each at a concentration of 100 μM). It is important not to use buffers containing primary amines such as Tris, as they will compete for conjugation with the amine reactive dyes. Probes are labeled as described in [18–20]. In short, 20 μg of probe mix containing 48 unlabeled *MDN1* probes, combined in an equimolar ratio, are dried down in a speed-vac and subsequently resuspended in 20 μl of 0.1 M sodium bicarbonate buffer (pH 9.0) containing monoreactive Cy3 dye that is sufficient to label 1 mg of protein or 30 nmol of amino groups. The mixture is then thoroughly vortexed and incubated

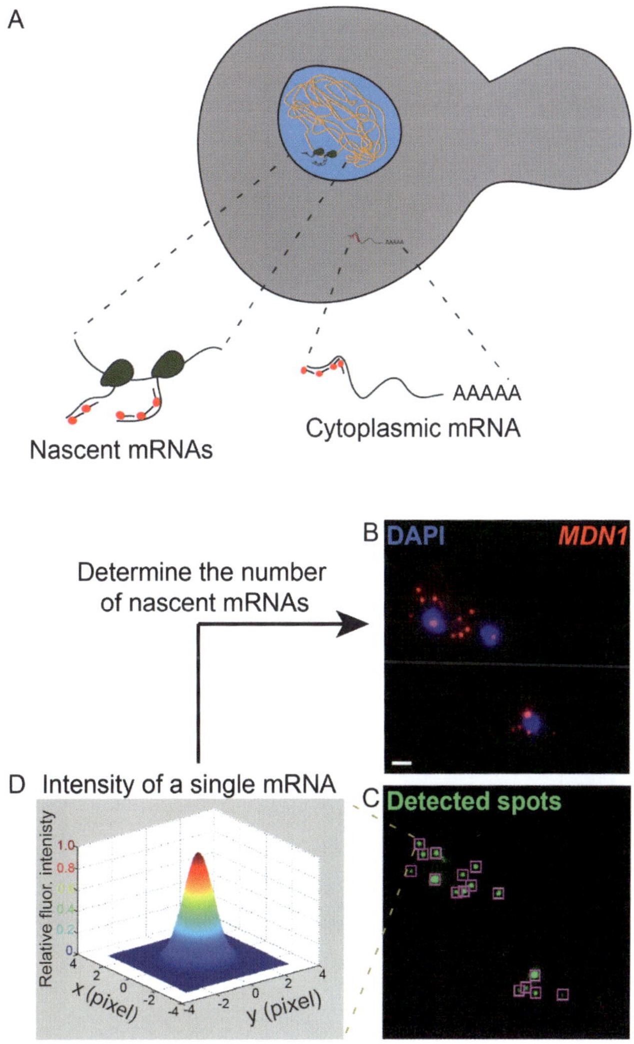

Fig. 1 (**a**) A schematic of smFISH probe binding to nascent and mature mRNAs. (**b**) Detection of cytoplasmic and nascent *MDN1* mRNAs (*red*) using smFISH in haploid W303 MATa cells. The nuclei are stained with a DAPI stain (*blue*). Note that transcription sites, located within the nuclei, are brighter than single cytoplasmic mRNAs indicating that several RNAP IIs are transcribing the *MDN1* gene concurrently. (**c**) A spot-detection algorithm was applied to detect cytoplasmic and nascent *MDN1* mRNAs. Detected mRNAs are marked as *green spots* demarcated by *purple squares*. (**d**) The intensity of a single cytoplasmic mRNA is determined. The total fluorescent intensity of a single spot in *x*, *y*, and *z* is determined. Once the average intensity of a single mRNA is known, it is used to calibrate the total intensity of transcription site to determine the absolute number of nascent chains associated with the active transcription sites detected in **b** and **c**. Scale bar: 1 μm

overnight at RT. Uncoupled Cy3 dye is removed using the QIAquick Nucleotide Removal Kit. Although labeling efficiency is often high (>75% of reactive amines are fluorescently coupled), unlabeled probes can be separated from the labeled probes by reverse phase high pressure liquid chromatography. Labeled probe mixes are diluted to the final concentration of 10 ng/μl and stored at −20 °C (*see* **Note 3**).

10. *Competitor DNA*. 100 μl of 10 mg/ml of competitor contains 50 μl of 10 mg/ml sheared salmon sperm DNA (Invitrogen) and 50 μl of 10 mg/ml *E. coli* tRNA (Roche). *E. coli* tRNA is prepared from solid by dissolving 10 mg in 1 ml of ddH_2O. S-tore at −20 °C.
11. *Hybridization wash solution*. 10% Formamide/2× SSC. Prepare fresh (*see* **Notes 4** and **5**). Use 2 ml per coverslip for every wash. Make a mastermix of 100 ml: 10 ml Formamide, 10 ml 20× SSC, 80 ml ddH_2O.
12. *Solution F.* 20% formamide, 10 mM $NaHPO_4$, pH 7.5. Prepare fresh. Use 12 μl per hybridization. Make a mastermix of 200 μl: 40 μl Formamide, 2 μl 1 M $NaHPO_4$ pH 7.5, 158 μl ddH_2O.
13. *Solution H.* 4× SSC, 2 mg/ml BSA, 10 mM VRC Prepare fresh. Use 12 μl per hybridization. Make a mastermix of 200 μl: 100 μl ddH_2O, 40 μl 20× SSC, 40 μl 10 mg/ml BSA, 20 μl 200 mM VRC.
14. *0.1% Triton X-100 solution*. To prepare 1 l of 0.1% Triton X-100 solution, combine 1 ml of 100% Triton X-100, 200 ml of 10× SSC and 799 ml of ddH_2O (*see* **Note 6**).
15. *4′,6-diamidino-2-phenylindole, dilactate (DAPI, dilactate)* (Sigma). Prepare a 0.5 μg/ml DAPI solution dissolved in 1× PBS in a Pyrex glass flask. Stir over night at RT. Wrap the flask in aluminum foil to protect from light. Store at 4 °C.
16. *Phosphate-buffered saline (PBS) buffer*. To prepare 1 l of 1× PBS, combine 800 ml of ddH_2O, 8 g of NaCl, 0.2 g of KCl, 1.44 g of Na_2HPO_4, and 0.24 g of KH_2PO_4. Adjust the pH to 7.4 with HCl and add additional ddH_2O to a total volume of 1 l.
17. *Microscope*. Images of smFISH-stained yeast cells are acquired with a widefield epifluorescence microscope, such as the Zeiss Axioimager Z2 equipped with an AxioCam mRm CCD camera. To maximize the number of detected photons, the sample is imaged with a high numerical aperture (NA) objective, such as a Zeiss Plan-Apochromat 100×/1.40 Oil objective with a working distance of 0.17 mm. A Zeiss immersion oil Immersol 518 F is used during imaging. The microscope was further equipped with the following filter sets: Chroma SP102 v1 (Cy3) and Chroma 31,000 (DAPI). The DC-powered

mercury lamp X-Cite® *exacte* (Excelitas Technologies) was used as a light source.

18. *Deconvolution algorithm.* Deconvolution restores three-dimensionally acquired (3D) images, which are affected by blurring and noise during imaging. Deconvolution increases the signal-to-noise (S/N) ratio and as such an effective resolution of an image. Increased S/N also eases detection and counting of single, smFISH hybridized mRNAs. A variety of commercially available deconvolution programs such Huygens deconvolution program provided by the Scientific Volume Imaging, though freely available programs can perform equally well [10, 19].
19. *Single molecule detection algorithm.* In budding yeast, mRNAs hybridized with smFISH probes appear as bright discrete fluorescent puncta distributed throughout the cytoplasm and the nucleus. To quantify the absolute number of transcripts and nascent chains per cell, a spot detection algorithm is employed capable of determining the number, fluorescent intensity and the position of each fluorescently labeled mRNA. Several freely available algorithms have been developed that are capable of quantifying smFISH-labeled mRNAs in two dimensions (2D) and in three dimensions (3D) [17–32]. Individual fluorescent spots are detected, their total fluorescent intensity determined and an average fluorescent insanity of a single cytoplasmic mRNA determined (Fig. 1b–d) [10, 18, 19, 31, 32].

3 Methods

3.1 Fixing and Permeabilizing Budding Yeast Cells

1. Grow 50 ml of yeast culture at 30 °C, shaking the flask at 250 rotations-per-minute (RPM) to accommodate exponential cell growth. When the OD_{600} reaches ~0.6, transfer the yeast into a 50 ml falcon tube and fix the cells by adding 7.1 ml of 32% paraformaldehyde. Mix immediately by inverting the falcon tube several times. Incubate at RT for 45 min and invert every 10 min.
2. Spin the cells at 2400 × *g* for 5 min at RT, discard the supernatant and wash pelleted cells three times with 10 ml of ice cold 1× buffer B, spinning the cells at 2400 × *g* for 5 min at 4 °C after each wash.
3. Resuspend the cells in 1 ml of spheroplast buffer and 800 U of lyticase and incubate for 10 min at 30 °C. Invert frequently. Spin for 4 min at 1300 × *g* at 4 °C and wash once in 1 ml of ice-cold buffer B. Resuspend in 1 ml 1× buffer B. Drop ~150 μl of cell suspension onto the coated coverslips placed in a 6- or 12-well tissue culture plate and incubate at 4 °C for

30 min to adhere the cells onto coverslips. Remove unattached cells with 5 ml of ice cold 1× buffer B, aspirate and store in 5 ml of 70% ethanol. Store at −20 °C until use (*see* **Note 7**).

3.2 Hybridization with smFISH Probes

1. Rehydrate the coverslips with adhered yeast cells in a Coplin jar filled with 2× SSC for 5 min at RT. Repeat the wash. Discard the solution and incubate in 5 ml of hybridization solution for 15 min at RT.
2. Set up hybridization mix containing smFISH probes. Per coverslip, combine 10 ng probe mix and 4 μl of 10 mg/ml competitor DNA and dry in a vacufuge (*see* **Note 3**). Resuspend the dried probe and competitor in 12 μl of solution F. Denature probe mix at 95 °C for 3 min, let cool slowly in the dark and then add 12 μl of solution H. Mix thoroughly and drop 20 μl into a petri dish. Take the coverslip, remove excess hybridization solution and place it face down onto the drop containing smFISH probes. Soak a Kimwipes paper with hybridization solution and place it in the petri dish away from the coverslips to humidify the hybridization chamber. Cover the petri dish, seal with Parafilm and incubate in 37 °C incubator for 3 h (*see* **Notes 4**, **5**, and **8**).
3. Place the coverslips from the petri dish into the Coplin jar with hybridization solution prewarmed to 37 °C to wash away unbound smFISH probes. Incubate for 20 min at 37 °C in dark. Repeat the wash (*see* **Note 9**).
4. Optional. Discard the hybridization solution and wash the yeast cells attached to the coverslips with 5 ml of 0.1% Triton X-100 solution for 15 min at RT while lightly shaking on the orbital shaker (*see* **Notes 6** and **9**). If Triton X-100 is omitted, proceed to **step 5**.
5. Wash twice with 1× PBS for 5 min at RT while lightly shaking on the orbital shaker. Incubate in a DAPI solution for 10 s and afterward wash with 1× PBS for 5 min.
6. Remove the coverslips from the Coplin jar, remove excess 1× PBS solution and mount the coverslips on to the microscope slide using a mounting medium. Seal the coverslips with nail polish (*see* **Notes 10** and **11**).

3.3 Imaging of the smFISH-Hybridized Yeast Cells

1. Using a widefield epifluorescence microscope (*see* Subheading 2), image the entire volume of a budding yeast cell using a 200 nm Z step (spanning approximately a 4.0 μm Z depth). Short exposure times are typically used to prevent excessive bleaching of the sample (*see* **Notes 3–5**, **12**, and **13**).
2. Optional. Deconvolve 3D images.

3.4 Single Molecule Detection and Counting

1. smFISH-labeled mRNAs are detected and counted on a maximally projected image (analyzing a two-dimensional (2D) image), or in a three dimensional (3D) image using a spot detection algorithm (*see* Subheading 2) [10, 19, 20, 23, 31, 32] (*see* **Note 14**).
2. To determine the average total fluorescent intensity of a single mRNA, the total fluorescent intensities of all mRNAs detected in an image are first plotted as a histogram. Afterward the distribution of these fluorescent intensities represented within a histogram is fitted to a Gaussian curve where the peak of the Gaussian fit represents the average total fluorescent intensity of a single mRNA detected in an image [19, 20, 32]. Cytoplasmic mRNAs can also group in a subcellular position to form mRNA clusters [19, 27, 32] (*see* **Note 15**). To determine the absolute number of mRNAs within a cluster, the total fluorescent intensity of each cluster is normalized by the average fluorescent intensity of a single mRNA. Clusters with a fluorescent intensity ranging between 0.5 and 1.5 are assumed to contain one smFISH-labeled transcript, while those with fluorescent intensities ranging between 1.5 and 2.5 are assumed to contain two smFISH-labeled transcripts and so on. Cells are then segmented and the absolute number of transcripts per cell is then determined [10, 17, 19, 20, 31, 32] (*see* **Notes 14** and **16**).
3. Determine the number of nascent mRNAs associated with an active site of transcription (*see* **Note 17**). Typically, active transcription sites are simultaneously transcribed by multiple RNAP II and thus an active transcription site will contain multiple nascent mRNAs. Each nascent chain will be hybridized with multiple smFISH probes thus appearing as large and bright nuclear fluorescent spots (Fig. 1b). To determine the absolute number of nascent mRNAs, the total fluorescent intensity of each transcription site is normalized by the average fluorescent intensity of a single mRNA. Transcription sites with a fluorescent intensity ranging between 0.5 and 1.5 are assumed to contain one smFISH-labeled nascent mRNA, those with fluorescent intensities ranging between 1.5 and 2.5 are assumed to contain two smFISH-labeled nascent mRNA and so on [10, 17, 31–34] (*see* **Notes 14–17**).

3.5 Quantifying mRNA Decay Rates Using smFISH

1. A mathematical model is used to quantify mRNA turnover rates. In any cell, cellular mRNA levels depend on the rate of their synthesis and the rate of their decay. Our mathematical model describes mRNA levels that are achieved as a balance between mRNA synthesis that follows zero-order kinetics and mRNA decay that follows first-order kinetics. We assume first-order mRNA decay kinetics because it is the simplest model that describes our smFISH-obtained data and can be used to compare our results with those obtained from the population

measurements of mRNA turnover [10, 17]. Using smFISH, total cellular mRNA levels and the rate of mRNA transcription can be quantified and thus the decay rate constant k_d (per unit of time) can be determined using the following differential equation:

$$\frac{dN}{dt} = \frac{m}{T} - k_d N, \tag{4}$$

where the first term of the equation describes the rate of mRNA synthesis (mRNAs per unit of time) and the second term describes the rate of mRNA decay (per unit of time). The equation has the following parameters: N is the absolute number of cellular mRNAs (*see* **Notes 14–16**); t is time and is determined with respect to the yeast cell cycle (*see* **Note 18**); m is the number of nascent transcripts associated with the active site of transcription and is determined experimentally; T is the average dwell time of a single nascent mRNA (*see* **Notes 19** and **20**). T further depends on the velocity of RNAP II, v (bases per unit of time), the length of the mRNA, l (bases), including the 5′ and the 3′ UTR, and the cleavage and release time of the nascent mRNA. For short genes, the cleavage and release time, also called the 3′ processing time, can be a substantial fraction of the total dwell time [35, 36]. A number of groups have experimentally determined the velocity of a transcribing RNAP II and the 3′ processing time [35, 37], while the length of the mRNA can be obtained from the *Saccharomyces genome database*. Thus,

$$\frac{dN}{dt} \approx \frac{mv}{l} - k_d N \tag{5}$$

When steady state is reached, the rate of mRNA synthesis is equal to the rate of mRNA decay, and the time-dependent changes in the number of cytoplasmic mRNAs become negligible. In this case, the number of cellular mRNAs at steady state (N_{ss}) can be described using the following equation:

$$N_{ss} = \frac{mv}{k_d l}. \tag{6}$$

Using this approach, we have calculated the mRNA half-life for *MDN1*, a constitutively expressed, housekeeping gene. In exponentially growing, haploid WT yeast cells, we have measured that at steady state, $N_{ss} = 4.65 \pm 1.2$ mRNAs and $m = 1.75 \pm 0.35$ nascent chains (Fig. 2a, b). These measurements were in a good agreement with our published reports [31]. $l = 14{,}733$ bases (available at Saccharomyces genome database) and $v = 2.0$ kb/min [37]. Assuming steady state conditions and using eq. 6, we derived a half-life of 13.5 min. The published $t_{1/2}$ for *MDN1* obtained by inhibiting

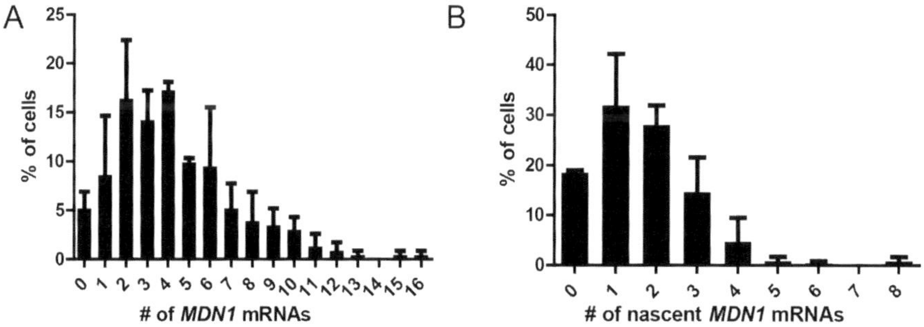

Fig. 2 (**a**) Distribution of mature *MDN1* mRNAs per cell. On average, a single cell contained 4.65 ± 1.2 *MDN1* transcripts (N_{eq}). (**b**) Distribution of nascent *MDN1* mRNAs per cell. On average, a single cell contained 1.75 ± 0.35 nascent *MDN1* transcripts (*m*), in agreement with previously published reports [31]

transcription in temperature sensitive mutants of the core transcriptional machinery is 26 min [13]. Normalization and transcription inhibition using a temperature sensitive RNAP II mutant *rpb1–1* likely prolonged the measured mRNA half-life by as much as a factor of two [9, 10, 34], which explains the discrepancy between the two methods and also highlights the importance of quantifying mRNA turnover in cells that have undergone minimal chemical or genetic perturbation (*see* **Notes 19–21**).

4 Notes

1. As a proof of principle, we use *MDN1* mRNA to quantify its mRNA half-life using smFISH and mathematical modeling. *MDN1* is the longest intronless gene in *S. cerevisiae* (when ranked on *Saccharomyces* genome database by the gene length) and can accommodate binding of 48 probes at the very 5′ end of the transcript, the maximal number of distinct smFISH Stellaris® probes allotted per purchased probe mix. Most genes however are shorter than *MDN1* and it is recommended that fewer probes are used to quantify transcriptional kinetics of those genes using the approach described here. When terminally labeled with a single fluorophore, a set containing ~25 probes annealing to different positions along the 5′ UTR should be used as this will lead to a more accurate quantification of the absolute number of nascent mRNAs, while still allowing single mRNA detection (unpublished observations). As an alternative, one can reliably quantify the absolute number of nascent chains associated with a single 50-nucleotide-long smFISH probe labeled with five Cy3 dyes, which anneals to the 5′ end of a nascent transcript [10, 33, 38]. Good results have also been obtained with four such probes, annealing to the very

5′ end of the transcript [10, 19, 20, 31, 33, 39, 40], however, synthesis of such 50-nucleotide-long probes is more expensive (*see* also **Note 4**).

2. In this work, commercially available, unlabeled and terminally reactive smFISH probes were used that were afterward coupled to a fluorescent Cy3 dye [18–20]. Alternatively, custom-designed and commercially available smFISH probes that are already fluorescently labeled and purified can also be used. A variety of fluorescent dyes are available (Biosearch Technologies, Inc.; www.biosearchtech.com/stellarisdesigner), where upon hybridization, CAL Fluor590 or Quasar 670 produce particularly strong fluorescent signals. Prelabeled smFISH probe mixes are shipped as dried mixes containing smFISH probes mixed in an equimolar ratio. These are then resuspended in TE buffer (defined below) to the final concentration of 10 ng/μl and stored at −20 °C (*see* **Note 3**). To make 100 ml of TE buffer (pH 8.0), combine 98.8 ml of ddH_2O with 1 ml of 1 M Tris–HCl (pH 8.0) and 0.2 ml of 0.5 M EDTA and adjust pH to 8.0. Store at R/T.

3. Fluorescent intensity is critical for differentiating between specific signal originating from single mRNAs hybridized with smFISH probes and background from nonspecifically bound by smFISH probes. Increasing the number of probes per mRNA increases signal-to-noise (S/N) ratio, detection sensitivity, and accuracy in transcript counting.

4. Hybridization conditions change depending on the length of smFISH probes. When 20-nucleotide-long probes are used, prehybridization and hybridization solutions are prepared with 10% formamide and 2× SSC (final concentrations). When 50-nucleotide-long probes are used, these solutions are prepared to contain 40% formamide and 2× SSC (final concentrations). In both cases, hybridization is then carried out at 37 °C. It should be noted that reducing formamide concentration and/or temperature will also result in more frequent nonspecific probe hybridization, while high formamide concentration and/or hybridization temperatures can prevent specific probe hybridization to the target mRNA. In both cases, the S/N will decrease, thereby decreasing the accuracy of single mRNA detection. Although standard conditions described here work for most probes, in some instances hybridization conditions might have to be optimized for specific probe sets.

5. Use only high quality deionized formamide. When exposed to air, deionized formamide quickly acidifies, which quenches fluorescence of dyes such as Quasar 670 during hybridization. To prevent acidification, aliquot formamide into microcentrifuge tubes and store at −20 °C for short-term storage and at

−80 °C for long term storage. To monitor the pH of deionized formamide, a pH strip can be used. Good quality deionized formamide has a pH between 7.5 and 8.

6. Triton X-100 helps to remove nonspecifically bound smFISH probes. However, prolonged incubation with Triton X-100 will result in loss of cytoplasmic content. Care should be taken to monitor the duration and the concentration of Triton X-100 during this wash.
7. For most *S. cerevisiae* strains, cell wall digestion occurs within the first 10 min of treatment with lyticase. Overdigestion will result in the loss of cellular morphology. Digestion should be monitored with a light microscope and bright field illumination; after ~90% of cells appear opaque, indicating that their wall was removed, the digestion should be stopped. After digestion, cells are sensitive to mechanical stress. Do not spin at high speeds or overpipette, otherwise dividing cells might lose their buds and change their morphology.
8. Hybridization can be carried out overnight and the rest of the protocol continued the next day.
9. Cover the Coplin jar with aluminum foil to protect smFISH probes in the sample from bleaching.
10. Certain commercial mounting media like ProLong Gold or ProLong Diamond must cure before samples can be imaged, while water based mounting media allow immediate imaging of the sample.
11. When stored at −20 °C, the smFISH-hybridized samples can be kept for several months.
12. Exposure times and the number of Z steps should be carefully set to minimize bleaching of smFISH probes during image acquisition. If bleaching is excessive, reduce exposure times and/or the number of Z steps. To increase the fluorescent intensity of a single mRNA it is better to increase the number of probes hybridizing to an mRNA rather than increasing the exposure time during imaging. With more probes used per probe mix, the exposure times will decrease as the fluorescent intensity of the mRNA will increase.
13. Single mRNA detection is sensitive to photobleaching. Use photostable dyes to label smFISH probes. We mostly use probes coupled with CAL Fluor590, Cy3, Quasar 670 or Cy5. Quasar 670 is bright but can rapidly bleach. It is sensitive to pH and its fluorescence can degrade when formamide becomes acidified (*see* **Note 5**).
14. mRNA numbers can be determined using different approaches. Reducing a 3D to a 2D dataset facilitates data analysis. One way of achieving this is by maximum image

projection, where only the pixel with the maximum intensity of all pixels in z at a specific *x*, *y* coordinate is projected to create a new 2D image. This works well for mRNAs that are low in abundance (~less than 40 mRNAs per cell) or not enriched in a small volume (clustered), as it is unlikely that multiple mRNAs are present in the exact same z position. For such mRNAs, single transcript counting is accurate by applying a spot detection algorithm on a maximal projected image. However, for highly abundant mRNAs, such as *ACT1*, or clustered mRNA, such as *ASH1*, mRNA levels should be quantified using 3D datasets [10, 17, 19, 21–32].

15. Single transcripts of clustered mRNAs like *ASH1*, which localizes to the bud tip, cannot be individually resolved [27]. To determine how many mRNAs reside in a cluster, a total fluorescent intensity of an individual cluster is normalized by the total fluorescent intensity of a single mRNA residing outside of the cluster [19, 27, 32]. Transcripts within a cluster and single transcripts found outside of the cluster are then summed to determine the absolute number of mRNAs per cells.
16. Threshold parameters will determine which fluorescent spots will be included as single mRNAs or single probes and optimizing threshold parameters is crucial for achieving single mRNA sensitivity. If the threshold is set too high, the algorithm will detect only the brightest fluorescent spots. If it is set too low, pixels with higher background fluorescence will also be detected, contributing to nonspecific noisy signals.
17. Majority of eukaryotic genes are composed of coding regions (exons) interjected by noncoding (intronic) regions. An engaged RNAP II transcribes exonic and intronic regions into an immature pre-mRNA, which is then spliced, resulting in a mature mRNA. These mature mRNAs are then exported through the nuclear pores into the cytoplasm. In mammalian cells, active site of transcription can be easily identified by using spectrally distinct smFISH probes that specifically hybridize to either intronic or exonic regions [17, 24, 31–34, 41, 42]. Most budding yeast genes however do not have introns and to differentiate nascent mRNAs from mature nucleoplasmic mRNAs that have left the site of transcription, a fluorescent intensity-based approach is used. Typically, when a gene is transcribed, multiple RNAP IIs are loaded onto the gene each simultaneously transcribing a pre-mRNA; active transcription sites are therefore typically of higher signal intensity than a single cytoplasmic mRNA or a mature nuclear transcript (Fig. 1b) [10, 17, 24, 30–34, 39–45]. Because multiple pre-mRNAs are being made, they will each allow hybridization of several smFISH probes. Thus, the total fluorescent intensity of an active transcription site is markedly enhanced relative to

the fluorescent intensity of a single mRNA (Fig. 1b). Thus, by designing smFISH probes that will hybridize to the 5′ most end of the mRNA, active sites of transcription can be identified as the brightest nuclear signal [10, 17, 24, 30–34, 39–45]. However, this is difficult to achieve for short genes transcribed at low frequency, as they may often have no, or only a single nascent mRNA. Nonetheless, nascent mRNAs can still be measured for these genes by quantifying the frequency of observing nuclear mRNA signals. mRNA export was shown to be fast and therefore, most nuclear signals of low expressed genes likely represents nascent mRNAs. Despite this uncertainty, mRNA degradation rates can be determined for infrequently expressed genes using the approach described in this protocol.

18. In this chapter, we describe a protocol that assumes a single mRNA decay rate that is invariant of the cell cycle progression. However, many mRNAs exhibit different stabilities depending on the cell cycle phase [10, 46–49]. To address this possibility using smFISH, a modified mathematical model is implemented. For a detailed explanation of this method, please refer to [10].
19. In this protocol we describe a method that assumes a single mRNA synthesis rate that is invariant of the cell cycle progression. A number of publications have quantified transcription rates in eukaryotic cells [35, 50], which can be used to determine mRNA half-lives using the smFISH approach and mathematical modeling described in this paper. However, transcriptional activity, as characterized by both the rate of RNAP IIs transcribing a gene and by the velocity of RNAP II transcribing the gene, can vary depending on the cell cycle phase [10, 35]. To address this possibility using smFISH, a modified mathematical model is implemented. For a detailed explanation of this method, please refer to [10].
20. To calculate mRNA half-lives using smFISH as described in this protocol, an average number of mRNAs per cell and average number of mRNAs associated with a transcription site are considered in the calculation. It is important to include all cells in the analysis, including those that have cytoplasmic transcripts but no detectable transcription site ($m = 0$), those that have a visible transcription site but no cytoplasmic transcript ($N = 0$), as well as those that have neither a visible transcription site nor detectable cytoplasmic transcripts. These cells also represent gene expression dynamics of a particular gene and omitting them in the analysis will alter the calculated mRNA decay rate.

21. When mRNA decay rates of an mRNA, whose expression oscillates through the cell cycle are investigated, the steady state conditions in Eqs. 5 and 6 cannot be assumed. In this case, parameter t, which is the length of each cell cycle phase (unit of time), as well as N and m for each cell cycle phase must be determined. For a detailed description of this approach, *see* [10].

Acknowledgments

We thank Daniel R. Larson for helpful comments on our manuscript. T.T. is an HHMI fellow of the Jane Coffin Childs Memorial Fund. D.Z. is supported by the Canadian Institute for Health Research (Project Grant-366682), Natural Sciences and Engineering Research Council of Canada and the Canadian Foundation for Innovation.

Author contributions: T.T. wrote the manuscript. S.R. and D.Z. provided the figures, and all three authors worked on the final design of the figures and manuscript text. All the authors approved the final version of the manuscript. The authors declare no competing financial interests.

References

1. Parker R, Sheth U (2007) P bodies and the control of mRNA translation and degradation. Mol Cell 25(5):635–646. https://doi.org/10.1016/j.molcel.2007.02.011
2. Herrick D, Parker R, Jacobson A (1990) Identification and comparison of stable and unstable mRNAs in Saccharomyces cerevisiae. Mol Cell Biol 10(5):2269–2284
3. Ross J (1995) mRNA stability in mammalian cells. Microbiol Rev 59(3):423–450
4. Wang Y, Liu CL, Storey JD, Tibshirani RJ, Herschlag D, Brown PO (2002) Precision and functional specificity in mRNA decay. Proc Natl Acad Sci U S A 99(9):5860–5865. https://doi.org/10.1073/pnas.092538799
5. Dodson RE, Shapiro DJ (2002) Regulation of pathways of mRNA destabilization and stabilization. Prog Nucleic Acid Res Mol Biol 72:129–164
6. Parker R, Herrick D, Peltz SW, Jacobson A (1991) Measurement of mRNA decay rates in Saccharomyces cerevisiae. Methods Enzymol 194:415–423
7. Passos DO, Parker R (2008) Analysis of cytoplasmic mRNA decay in Saccharomyces cerevisiae. Methods Enzymol 448:409–427. https://doi.org/10.1016/S0076-6879(08)02620-7
8. Miller C, Schwalb B, Maier K, Schulz D, Dumcke S, Zacher B, Mayer A, Sydow J, Marcinowski L, Dolken L, Martin DE, Tresch A, Cramer P (2011) Dynamic transcriptome analysis measures rates of mRNA synthesis and decay in yeast. Mol Syst Biol 7:458. https://doi.org/10.1038/msb.2010.112
9. Sun M, Schwalb B, Schulz D, Pirkl N, Etzold S, Lariviere L, Maier KC, Seizl M, Tresch A, Cramer P (2012) Comparative dynamic transcriptome analysis (cDTA) reveals mutual feedback between mRNA synthesis and degradation. Genome Res 22(7):1350–1359. https://doi.org/10.1101/gr.130161.111
10. Trcek T, Larson DR, Moldon A, Query CC, Singer RH (2011) Single-molecule mRNA decay measurements reveal promoter-regulated mRNA stability in yeast. Cell 147(7):1484–1497. https://doi.org/10.1016/j.cell.2011.11.051
11. Pelechano V, Perez-Ortin JE (2008) The transcriptional inhibitor thiolutin blocks mRNA degradation in yeast. Yeast 25(2):85–92. https://doi.org/10.1002/yea.1548

12. Grigull J, Mnaimneh S, Pootoolal J, Robinson MD, Hughes TR (2004) Genome-wide analysis of mRNA stability using transcription inhibitors and microarrays reveals posttranscriptional control of ribosome biogenesis factors. Mol Cell Biol 24 (12):5534–5547. https://doi.org/10.1128/MCB.24.12.5534-5547.2004
13. Holstege FC, Jennings EG, Wyrick JJ, Lee TI, Hengartner CJ, Green MR, Golub TR, Lander ES, Young RA (1998) Dissecting the regulatory circuitry of a eukaryotic genome. Cell 95(5):717–728
14. Livak KJ, Schmittgen TD (2001) Analysis of relative gene expression data using real-time quantitative PCR and the 2(−Delta Delta C (T)) Method. Methods 25(4):402–408. https://doi.org/10.1006/meth.2001.1262
15. Bregman A, Avraham-Kelbert M, Barkai O, Duek L, Guterman A, Choder M (2011) Promoter elements regulate cytoplasmic mRNA decay. Cell 147(7):1473–1483. https://doi.org/10.1016/j.cell.2011.12.005
16. Enssle J, Kugler W, Hentze MW, Kulozik AE (1993) Determination of mRNA fate by different RNA polymerase II promoters. Proc Natl Acad Sci U S A 90(21):10091–10095
17. Bahar Halpern K, Itzkovitz S (2016) Single molecule approaches for quantifying transcription and degradation rates in intact mammalian tissues. Methods 98:134–142. https://doi.org/10.1016/j.ymeth.2015.11.015
18. Castelnuovo M, Rahman S, Guffanti E, Infantino V, Stutz F, Zenklusen D (2013) Bimodal expression of PHO84 is modulated by early termination of antisense transcription. Nat Struct Mol Biol 20(7):851–858. https://doi.org/10.1038/nsmb.2598
19. Trcek T, Chao JA, Larson DR, Park HY, Zenklusen D, Shenoy SM, Singer RH (2012) Single-mRNA counting using fluorescent in situ hybridization in budding yeast. Nat Protoc 7 (2):408–419. https://doi.org/10.1038/nprot.2011.451
20. Zenklusen D, Singer RH (2010) Analyzing mRNA expression using single mRNA resolution fluorescent in situ hybridization. Methods Enzymol 470:641–659. https://doi.org/10.1016/S0076-6879(10)70026-4
21. Battich N, Stoeger T, Pelkmans L (2013) Image-based transcriptomics in thousands of single human cells at single-molecule resolution. Nat Methods 10(11):1127–1133. https://doi.org/10.1038/nmeth.2657
22. Chen KH, Boettiger AN, Moffitt JR, Wang S, Zhuang X (2015) RNA imaging. Spatially resolved, highly multiplexed RNA profiling in single cells. Science 348(6233):aaa6090. https://doi.org/10.1126/science.aaa6090
23. Lionnet T, Czaplinski K, Darzacq X, Shav-Tal Y, Wells AL, Chao JA, Park HY, de Turris V, Lopez-Jones M, Singer RH (2011) A transgenic mouse for in vivo detection of endogenous labeled mRNA. Nat Methods 8 (2):165–170. https://doi.org/10.1038/nmeth.1551
24. Raj A, Peskin CS, Tranchina D, Vargas DY, Tyagi S (2006) Stochastic mRNA synthesis in mammalian cells. PLoS Biol 4(10):e309. https://doi.org/10.1371/journal.pbio.0040309
25. Raj A, van den Bogaard P, Rifkin SA, van Oudenaarden A, Tyagi S (2008) Imaging individual mRNA molecules using multiple singly labeled probes. Nat Methods 5(10):877–879. https://doi.org/10.1038/nmeth.1253
26. Skinner SO, Sepulveda LA, Xu H, Golding I (2013) Measuring mRNA copy number in individual Escherichia coli cells using single-molecule fluorescent in situ hybridization. Nat Protoc 8(6):1100–1113. https://doi.org/10.1038/nprot.2013.066
27. Trcek T, Grosch M, York A, Shroff H, Lionnet T, Lehmann R (2015) Drosophila germ granules are structured and contain homotypic mRNA clusters. Nat Commun 6:7962. https://doi.org/10.1038/ncomms8962
28. Trovisco V, Belaya K, Nashchekin D, Irion U, Sirinakis G, Butler R, Lee JJ, Gavis ER, St Johnston D (2016) Bicoid mRNA localises to the Drosophila oocyte anterior by random dynein-mediated transport and anchoring. elife 5:e17537. https://doi.org/10.7554/eLife.17537
29. Vargas DY, Shah K, Batish M, Levandoski M, Sinha S, Marras SA, Schedl P, Tyagi S (2011) Single-molecule imaging of transcriptionally coupled and uncoupled splicing. Cell 147 (5):1054–1065. https://doi.org/10.1016/j.cell.2011.10.024
30. Xu H, Sepulveda LA, Figard L, Sokac AM, Golding I (2015) Combining protein and mRNA quantification to decipher transcriptional regulation. Nat Methods 12 (8):739–742. https://doi.org/10.1038/nmeth.3446
31. Zenklusen D, Larson DR, Singer RH (2008) Single-RNA counting reveals alternative modes of gene expression in yeast. Nat Struct Mol Biol 15(12):1263–1271. https://doi.org/10.1038/nsmb.1514
32. Trcek T, Lionnet T, Shroff H, Lehmann R (2017) mRNA quantification using single-molecule FISH in Drosophila embryos. Nat

Protoc 12(7):1326–1348. https://doi.org/10.1038/nprot.2017.030
33. Femino AM, Fay FS, Fogarty K, Singer RH (1998) Visualization of single RNA transcripts in situ. Science 280(5363):585–590
34. Gandhi SJ, Zenklusen D, Lionnet T, Singer RH (2011) Transcription of functionally related constitutive genes is not coordinated. Nat Struct Mol Biol 18(1):27–34. https://doi.org/10.1038/nsmb.1934
35. Larson DR, Zenklusen D, Wu B, Chao JA, Singer RH (2011) Real-time observation of transcription initiation and elongation on an endogenous yeast gene. Science 332 (6028):475–478. https://doi.org/10.1126/science.1202142
36. Palangat M, Larson DR (2012) Complexity of RNA polymerase II elongation dynamics. Biochim Biophys Acta 1819(7):667–672. https://doi.org/10.1016/j.bbagrm.2012.02.024
37. Mason PB, Struhl K (2005) Distinction and relationship between elongation rate and processivity of RNA polymerase II in vivo. Mol Cell 17(6):831–840. https://doi.org/10.1016/j.molcel.2005.02.017
38. Chubb JR, Trcek T, Shenoy SM, Singer RH (2006) Transcriptional pulsing of a developmental gene. Curr Biol 16(10):1018–1025. https://doi.org/10.1016/j.cub.2006.03.092
39. Levsky JM, Shenoy SM, Pezo RC, Singer RH (2002) Single-cell gene expression profiling. Science 297(5582):836–840. https://doi.org/10.1126/science.1072241
40. Pezo RC, Gandhi SJ, Shirley LA, Pestell RG, Augenlicht LH, Singer RH (2008) Single-cell transcription site activation predicts chemotherapy response in human colorectal tumors. Cancer Res 68(13):4977–4982. https://doi.org/10.1158/0008-5472.CAN-07-6770
41. Cho WK, Jayanth N, English BP, Inoue T, Andrews JO, Conway W, Grimm JB, Spille JH, Lavis LD, Lionnet T, Cisse II (2016) RNA polymerase II cluster dynamics predict mRNA output in living cells. eLife 5. https://doi.org/10.7554/eLife.13617
42. Hoyle NP, Ish-Horowicz D (2013) Transcript processing and export kinetics are rate-limiting steps in expressing vertebrate segmentation clock genes. Proc Natl Acad Sci U S A 110 (46):E4316–E4324. https://doi.org/10.1073/pnas.1308811110
43. Lagha M, Bothma JP, Esposito E, Ng S, Stefanik L, Tsui C, Johnston J, Chen K, Gilmour DS, Zeitlinger J, Levine MS (2013) Paused Pol II coordinates tissue morphogenesis in the Drosophila embryo. Cell 153(5):976–987. https://doi.org/10.1016/j.cell.2013.04.045
44. Levesque MJ, Raj A (2013) Single-chromosome transcriptional profiling reveals chromosomal gene expression regulation. Nat Methods 10(3):246–248. https://doi.org/10.1038/nmeth.2372
45. Little SC, Tikhonov M, Gregor T (2013) Precise developmental gene expression arises from globally stochastic transcriptional activity. Cell 154(4):789–800. https://doi.org/10.1016/j.cell.2013.07.025
46. Marzluff WF, Wagner EJ, Duronio RJ (2008) Metabolism and regulation of canonical histone mRNAs: life without a poly(A) tail. Nat Rev Genet 9(11):843–854. https://doi.org/10.1038/nrg2438
47. Messier V, Zenklusen D, Michnick SW (2013) A nutrient-responsive pathway that determines M phase timing through control of B-cyclin mRNA stability. Cell 153(5):1080–1093. https://doi.org/10.1016/j.cell.2013.04.035
48. Osley MA (1991) The regulation of histone synthesis in the cell cycle. Annu Rev Biochem 60:827–861. https://doi.org/10.1146/annurev.bi.60.070191.004143
49. Talarek N, Cameroni E, Jaquenoud M, Luo X, Bontron S, Lippman S, Devgan G, Snyder M, Broach JR, De Virgilio C (2010) Initiation of the TORC1-regulated G0 program requires Igo1/2, which license specific mRNAs to evade degradation via the 5′–3′ mRNA decay pathway. Mol Cell 38(3):345–355. https://doi.org/10.1016/j.molcel.2010.02.039
50. Ardehali MB, Lis JT (2009) Tracking rates of transcription and splicing in vivo. Nat Struct Mol Biol 16(11):1123–1124. https://doi.org/10.1038/nsmb1109-1123

Chapter 5

PAR-CLIP for Discovering Target Sites of RNA-Binding Proteins

Aitor Garzia, Pavel Morozov, Marcin Sajek, Cindy Meyer, and Thomas Tuschl

Abstract

RNA-binding proteins (RBPs) establish posttranscriptional gene regulation (PTGR) by coordinating the maturation, editing, transport, stability, and translation of cellular RNAs. A variety of experimental approaches have been developed to characterize the RNAs associated with RBPs in vitro as well as in vivo. Our laboratory developed Photoactivatable-Ribonucleoside-Enhanced Cross-Linking and Immunoprecipitation (PAR-CLIP), which in combination with next-generation sequencing enables the identification of RNA targets of RBPs at a nucleotide-level resolution. Here we present an updated and condensed step-by-step PAR-CLIP protocol followed by the description of our RNA-seq data analysis pipeline.

Key words RNA recognition element, Next-generation sequencing, Ribonucleoprotein, PAR-CLIP Suite

1 Introduction

Regulation of the maturation, editing, transport, stability, and translation of all categories of cellular RNAs is collectively considered posttranscriptional gene regulation (PTGR). These processes are coordinated by RNA-binding proteins (RBPs) and ribonucleoprotein complexes (RNPs) [1–3]. In humans, RBPs comprise an abundant class of proteins interacting with at least one category of RNA, e.g., mRNA or tRNA. There are over 1500 RBPs corresponding to approximately 7.5% of all human protein-coding genes [4]. Each RBP contains one or more RNA-binding domains (RBDs), some of which specifically recognize target transcripts and bind at defined sequence and structural elements termed RNA recognition elements (RREs), while others are guided to specific subcellular compartments or RNA processing complexes for interacting with their targets. Most RBP families have high conservation and low tissue specificity, suggesting that PTGR processes support

Shireen R. Lamandé (ed.), *mRNA Decay: Methods and Protocols*, Methods in Molecular Biology, vol. 1720,
https://doi.org/10.1007/978-1-4939-7540-2_5,

basic cellular functions and are ancient and equally essential for all cells [4].

Characterizing interactions between an RBP of interest and its target RNAs is essential to dissect their functions and different RNA regulatory processes. The genome-wide identification of endogenous targets of RBPs has been facilitated by several high-throughput technologies, especially immunoprecipitation (IP) [5, 6] of proteins UV-cross-linked to RNAs [7, 8] in combination with next-generation sequencing and modern protein mass spectrometry [9–11]. The application of those techniques revealed that many RBPs bind to thousands of transcripts in living cells at defined binding sites. However, one major drawback is the presence of high background signal caused by copurified non-cross-linked cellular RNAs or RBPs and their targets. To be able to separate cross-linked target RNAs from copurified background nucleic acids, our laboratory developed Photoactivatable-Ribonucleoside-Enhanced Cross-Linking and Immunoprecipitation (PAR-CLIP) [12].

In a PAR-CLIP experiment, highly photoreactive ribonucleosides, such as 4-thiouridine (4SU), are incorporated into newly transcribed RNAs in living cells. After cross-linking, the RBP of interest is immunoprecipitated, the bound RNAs are isolated, and converted into a cDNA library. During reverse transcription, the reverse transcriptase enzyme introduces a characteristic mutation at the position of the cross-linked nucleotide (T-to-C for 4SU). Those characteristic transitions allow the separation of cross-linked from coisolated non-cross-linked input RNAs and facilitate the computational identification of the underlying RNA recognition element (RRE).

PAR-CLIP has been applied successfully not only to study over 50 human RBP families in human cell lines [12–16], but also in cell lines from other species, including *Drosophila melanogaster* [17, 18] and mice [14, 15] as well as in entire organisms, such as *Caenorhabditis elegans* [19, 20] and *Saccharomyces cerevisiae* [21, 22].

Recently our laboratory published a detailed and revised protocol with explanations of the critical steps the experimentalist should consider before performing a PAR-CLIP experiment, as well as detailed description of our computational analysis pipeline that determines the quality of PAR-CLIP cDNA libraries [23]. The aim of this chapter is to present a concise protocol for the experimentalist though it will also refer to recent sources for more detail.

2 Materials

1. 1 M 4-Thiouridine (4SU) stock solution: 260.27 mg 4SU in 1 ml DMSO.
2. 10 mg/ml Doxycycline stock solution: 10 mg doxycycline in 1 ml DMSO.

3. PBS buffer: 137 mM NaCl, 2.7 mM KCl, 10 mM Na_2HPO_4, 1.8 mM KH_2PO_4.
4. PBS-T buffer: PBS supplied with 0.1% Tween 20.
5. 1× NP40 lysis buffer: 50 mM HEPES, pH 7.5, 150 mM KCl, 2 mM EDTA, 2% (v/v) NP40, 0.5 mM DTT, complete EDTA-free Protease Inhibitor Cocktail (Roche), Phosphatase Inhibitor Cocktail tablets *PhosSTOP* (Roche).
6. IP wash buffer: 50 mM HEPES-KOH, pH 7.5, 300 mM KCl, 0.05% (v/v) NP40.
7. High salt wash buffer: 50 mM HEPES-KOH, pH 7.5, 500 mM KCl, 0.05% (v/v) NP40.
8. Dephosphorylation buffer: 50 mM Tris–HCl, pH 7.9, 100 mM NaCl, 10 mM $MgCl_2$, 1 mM DTT.
9. Phosphatase wash buffer: 50 mM Tris–HCl, pH 7.5, 20 mM EGTA, 0.5% (v/v) NP40.
10. Polynucleotide kinase (PNK) buffer without DTT: 50 mM Tris–HCl, pH 7.5, 50 mM NaCl, 10 mM $MgCl_2$.
11. PNK buffer with DTT: 50 mM Tris–HCl, pH 7.5, 50 mM NaCl, 10 mM $MgCl_2$, 5 mM DTT.
12. 4× SDS PAGE loading buffer: 10% glycerol (v/v), 50 mM Tris–HCl, pH 6.8, 2 mM EDTA, 2% SDS (w/v), 100 mM DTT, 0.1% Bromophenol blue.
13. 4× Proteinase K buffer: 200 mM Tris–HCl, pH 7.5, 300 mM NaCl, 25 mM EDTA, 4% (w/v) SDS.
14. Acidic phenol–chloroform–isoamyl alcohol: 25 ml acidic phenol, pH 4.2, 24 ml chloroform, 1 ml isoamyl alcohol.
15. 10× RNA ligase buffer without ATP: 500 mM Tris–HCl, pH 7.6, 100 mM $MgCl_2$, 100 mM 2-mercaptoethanol, 1 mg/ml acetylated BSA (Sigma).
16. 10× RNA ligase buffer with ATP: 500 mM Tris–HCl, pH 7.6, 100 mM $MgCl_2$, 100 mM 2-mercaptoethanol, 1 mg/ml acetylated BSA (Sigma), 2 mM ATP.
17. Denaturing PAA gel loading solution: 98.8% formamide, 1% (v/v) 0.5 M Na_2 H_2EDTA, pH 8.0, 0.2% Bromophenol blue.
18. 150 mM KOH/20 mM Tris base and 150 mM HCl. Use pH paper to verify that a 1:1 mixture of these two solutions results in a pH between 7.0 and 9.5.
19. 10× dNTP solution: 2 mM dATP, 2 mM dCTP, 2 mM dGTP, 2 mM dTTP.
20. 10× PCR buffer: 100 mM Tris–HCl pH 8.0, 500 mM KCl, 1% Triton X-100, 20 mM $MgCl_2$, 10 mM 2-mercaptoethanol.

21. Spectrolinker XL-1500 (Spectronics Corporation) equipped with 365 nm light bulbs.
22. DynaMag-2 and 15 magnets (ThermoFisher).
23. Model V15.17 Gel Electrophoresis Apparatus (Biometra).
24. Glass Plates and spacer sets for 15 cm wide, 17 cm long, 0.5 mm tick, 30 ml gel volume (Biometra).
25. 1 mg/ml ANTI-FLAG M2, mouse monoclonal (Sigma).
26. Dynabeads Protein G (ThermoFisher).
27. 50 ml centrifuge tubes.
28. 15 ml centrifuge tubes.
29. 2 ml siliconized microtubes (BioPlas).
30. 1.5 ml siliconized microtubes (BioPlas).
31. 0.2 μm membrane syringe filter (Pall Acrodisc).
32. 1000 U/μl RNase T1 (ThermoFisher).
33. 10 mg/ml RNase A (ThermoFisher).
34. 10 U/μl RNase I (ThermoFisher).
35. Calf intestinal phosphatase (CIP) (New England Biolabs).
36. T4 Polynucleotide Kinase (T4 PNK) (New England Biolabs).
37. 10 mCi/ml, 1.6 μM γ-^{32}P-ATP (PerkinElmer).
38. Protein size marker, Precision Plus Protein Standard (Bio-Rad).
39. NuPAGE Novex 4–12% Bis-Tris Protein Gels (ThermoFisher).
40. NuPAGE MOPS SDS Running Buffer (20×) (ThermoFisher).
41. D-Tube Dialyzer Midi, cutoff 3.5 kDa (Novagen).
42. Proteinase K (Roche).
43. 10 mg/ml Glycoblue or glycogen (Ambion or Roche).
44. 1 mg/ml truncated and mutated RNA ligase 2, T4 Rnl2 (1-249) K227Q (New England Biolabs or plasmid for recombinant expression can also be obtained at addgene.org).
45. 1 mg/ml T4 RNA ligase, T4 Rnl1 (ThermoFisher).
46. SuperScript III Reverse Transcriptase (ThermoFisher).
47. 5 U/μl *Taq* DNA polymerase.
48. Qiaquick gel purification kit (Qiagen).
49. RNA size marker 19 nt: 5′ CGUACGCGGGUUUAAACGA 3′.
50. RNA size marker 35 nt: 5′ CUCAUCUUGGUCGUACGCG-GAAUAGUUUAAACUGU 3′.
51. Preadenylated 3′ adapter (DNA): 5′ AppTCGTATGCCG TCTTCTGCTTGT 3′.

52. 5′ adapter (RNA): 5′ GUUCAGAGUUCUACAGUCCGAC-GAUC 3′.
53. 3′ primer: 5′ CAAGCAGAAGACGGCATACGA 3′.
54. 5′ primer: 5′ AATGATACGGCGACCACCGACAGGTTCA-GAGTTCTACAGTCCGA 3′.

3 Methods (*See* Notes 1 and 2)

3.1 Preparation of Lysate

3.1.1 Expanding Cells

1. Expand cells to approximately 80% confluency in appropriate growth medium in 150-mm tissue culture dishes. We recommend using approximately 100–200 × 10^6 cells or a number of cells that will result in 1.5–3 ml of wet pellet.
2. For tagged proteins expressed inducibly, add doxycycline to a final concentration of 1 μg/ml 24 h before cross-linking.
3. Sixteen hours before cross-linking add 4SU to a final concentration of 100 μM directly to the cell culture medium (*see* **Notes 3–5**).

3.1.2 UV365-Cross-Linking

1. Aspirate or pour off the media from tissue culture dishes.
2. Wash cells once with 5 ml ice-cold PBS per dish and remove PBS completely.
3. Place dishes on a tray with ice and irradiate uncovered with a dose of 0.15 J/cm^2 of 365 nm.
4. Scrape cells off with a cell scraper in 1 ml PBS per dish, transfer to 50 ml centrifugation tubes, collect by centrifugation at 500 × *g* at 4 °C for 15 min and discard the supernatant (*see* **Note 6**).
5. Wash cells once with 5 ml ice-cold PBS per 1 ml of wet cell pellet and collect by centrifugation at 500 × *g* at 4 °C for 15 min. Discard the supernatant.
6. Unless continuing directly with cell lysis, snap freeze the cell pellet in liquid nitrogen and store at −80 °C. Cell pellets can be stored for at least 12 months.

3.1.3 Cell Lysis and RNase Digestion

The RNase treatment needs to be optimized for each PAR-CLIP experiment. (*see* **Notes** 7 and **8**).

1. Resuspend cell pellet described under Subheading 3.1.2 in 3 volumes of 1× NP40 lysis buffer and incubate on ice for 30 min.
2. Centrifuge at 13,000 × *g* at 4 °C for 30 min. Collect the supernatant. In the meantime, prepare the magnetic beads (*see* below).

3. Add the corresponding RNase to the supernatant to an optimal final concentration (usually 1 U/μl) and incubate at room temperature for 15 min. Cool reaction subsequently for 5 min on ice before proceeding.

3.2 Immunoprecipitation and Recovery of Cross-Linked RNA

The IP efficiencies strongly depend on the antibody and magnetic beads that are used for IP (*see* **Notes 9–11**).

3.2.1 Preparation of Magnetic Beads

1. Transfer 10 μl of Protein G magnetic particles per ml cell lysate to a 2 ml microtube. Wash beads twice with 1 ml of PBS-T buffer.
2. Resuspend the beads in twice the volume of PBS-T buffer relative to the original volume of bead suspension.
3. Add 0.25 mg of anti-FLAG M2 antibody (1 mg/ml) per ml bead suspension and incubate on a rotating wheel for 60 min at room temperature.
4. Wash beads twice in 1 ml of PBS-T buffer to remove unbound antibody.
5. Resuspend beads in 1 volume of PBS-T buffer.

3.2.2 Immunoprecipitation (IP) and Dephosphorylation

1. Add 10 μl of freshly prepared antibody-conjugated magnetic beads per ml of partial RNase treated cell lysate described under Subheading 3.1.3, **step 3** and incubate in 15 ml centrifugation tubes on a rotating wheel for 2 h at 4 °C.
2. Collect magnetic beads on a magnetic particle collector for 15 centrifugation tubes and discard supernatant (*see* **Note 12**).
3. Add 1 ml of IP wash buffer and transfer to 2 ml microtubes (*see* **Notes 13** and **14**).
4. Wash beads three times in 1 ml of IP wash buffer.
5. Resuspend beads in one bead volume of IP wash buffer.
6. If necessary, perform second RNase digest (*see* **Note 15**). Add RNase to a final concentration of 1–100 U/μl to the bead suspension and incubate for 15 min at 22 °C.
7. Wash beads three times in 1 ml of high salt wash buffer (*see* **Note 13**).
8. Resuspend beads in 1 volume of dephosphorylation buffer (*see* **Note 16**).
9. Add calf intestinal alkaline phosphatase (10 U/μl) to a final concentration of 0.5 U/μl, and incubate the suspension for 15 min at 37 °C.
10. Wash beads twice in 1 ml of phosphatase wash buffer.

11. Wash beads twice in 1 ml PNK buffer without DTT.
12. Resuspend beads in one original bead volume of PNK buffer with DTT (DTT is necessary for enzymatic activity, 1× PNK buffer).

*3.2.3 Radiolabeling of Immunoprecipitated RNA (See **Note 17**)*

1. Add γ-^{32}P-ATP (0.01 mCi/μl) to a final concentration of 0.5 μCi/μl (1.6 μM ATP) and T4 PNK to a final concentration of 1 U/μl to the bead suspension described under Subheading 3.2.2, **step 12**. Incubate the suspension for 30 min at 37 °C.
2. Add nonradioactive ATP to obtain a final concentration of 100 μM and incubate for another 15 min at 37 °C.
3. Wash the magnetic beads five times with 1 ml of PNK buffer without DTT. Transfer beads to a fresh microtube after the second wash.
4. Resuspend the beads in 70 μl of 1× SDS-PAGE loading buffer.
5. Incubate the radiolabeled suspension for 5 min in a heat block at 95 °C to denature and release the immunoprecipitated proteins with cross-linked RNA.
6. Separate the magnetic beads on the magnet and transfer the supernatant to a clean 1.5 ml microtube.

*3.2.4 SDS-PAGE and Electroelution of Cross-Linked RNA–Protein Complexes from Gel Slices (See **Note 18**)*

1. Load 60 μl of the supernatant described under Subheading 3.2.3, **step 6** per well of a precast Novex Bis-Tris 4–12% polyacrylamide gel and run the gel for 1 h at 200 V. Keep 5 μl of supernatant for a Western blot to confirm that the IP was efficient. Space samples appropriately, typically at a two-well distance, to avoid cross-contamination.
2. Disassemble the gel chamber and dismantle the gel, leaving it mounted on one plate. To facilitate the alignment of the gel to the phosphorimager paper printout, we recommend implanting three tiny radioactive gel pieces asymmetrically at three of the four corners of the gel. Wrap the gel in plastic film (e.g., Saran wrap) to avoid contamination.
3. Expose the gel to a blanked phosphorimager screen for 1 h and visualize on a phosphorimager.
4. Align the gel on top of the phosphorimager printout using the implanted gel pieces for orientation. Excise the bands that correspond to the expected size of the RBPs.
5. Add 600 μl of H_2O to a D-Tube Dialyzer Midi Tube and let it stand at room temperature for 5 min. Discard the water.
6. Transfer the excised band to the dialyzer tube and add 600 μl 1× SDS running buffer (filtered by 0.2 μm membrane syringe filter).

7. Electroelute the cross-linked RNA–RBP complex in 1× SDS running buffer at 100 V for 2 h.
8. Transfer the supernatant (600 μl) to a fresh 1.5 ml microtube. To the excised gel bands add another 600 μl 1× SDS running buffer and electroelute for another 2 h.
9. Transfer the supernatant (600 μl) to a clean 1.5 ml microtube.

3.2.5 Proteinase K Digestion

1. Add 200 μl of 4× Proteinase K Buffer to each of the electroeluate-containing microtubes, followed by the addition of Proteinase K to a final concentration of 1.0 mg/ml. Incubate for 30 min at 50 °C.
2. Recover the RNA by acidic phenol–chloroform–isoamyl alcohol extraction (25:24:1, pH 4.0) followed by a chloroform extraction. Add NaCl to a final concentration of 0.3 M, followed by 1 μl of glycogen and 3 volumes of ethanol.
3. Allow the RNA to precipitate by incubation in the freezer for at least 30 min and collect the pellet by centrifugation in a centrifuge at 4 °C at maximum speed (approx. 14,000 × *g*) for 20 min.
4. Discard the supernatant. Wash the microtube wall and pellet with 1 ml 75% ethanol and centrifuge at 4 °C at maximum speed (approx. 14,000 × *g*) for 10 min.
5. Discard the supernatant. Collect residual ethanol by centrifugation at 14,000 × *g* for 1 min and discard the supernatant. Air-dry the RNA pellet for 5 min.
6. Dissolve all pellets from the same sample in 10 μl water.

3.3 cDNA Library Preparation

3.3.1 Preparation of Radioactive Size Markers

1. Radiolabel the size markers (19 and 35 nt) individually in a 10 μl reaction by combining in a siliconized microtube:
   - 1 μl of 10 μM oligoribonucleotide.
   - 1 μl of 10× PNK buffer.
   - 1 μl of γ-^{32}P-ATP (0.01 mCi/μl).
   - 6 μl of water.
2. Incubate sample at 90 °C for 1 min.
3. Incubate sample on ice for 2 min.
4. Add 1 μl of T4 PNK enzyme (10 U/μl).
5. Incubate sample for 30 min at 37 °C.
6. Quench the reaction by addition of 10 μl of denaturing PAA gel loading solution.
7. Incubate sample at 90 °C for 1 min.
8. Load 20 μl of each sample in a single well on a 15% denaturing acrylamide gel. Run the gel approx. 45 min at 30 W using 0.5×

TBE buffer until the Bromophenol blue band is close to the bottom of the gel.

9. Dismantle the gel, leaving it mounted on one glass plate. Implant three tiny radioactive gel pieces asymmetrically at three of the four corners of the gel, to facilitate the alignment of the gel to the phosphorimager paper printout. Wrap the gel in plastic film (e.g., Saran wrap) to avoid contamination. Expose the gel for 1 min and image the gel on a phosphorimager.
10. Align the gel on glass plate over the printed phosphoimage. Cut out the radioactive bands corresponding to the size marker and transfer the gel slices into preweighed 1.5 ml siliconized microtubes (one for each marker).
11. Add 400 μl of 0.3 M NaCl to each microtube and elute the RNA from the gel by incubating the microtube for at least 1 h at 37 °C under constant agitation at 1000 rpm.
12. Collect the supernatant and add 3 volumes of ethanol of the volume of 0.3 M NaCl used in **step 11**. Keep sample at −20 °C for at least 30 min.
13. Collect the RNA by centrifugation in tabletop centrifuge at 4 °C at maximum speed (approx. 14,000 × *g*) for 15 min.
14. Discard the supernatant. Collect residual ethanol by centrifugation at 14,000 × *g* for 10 s. Air-dry the RNA pellet for 5 min.
15. Dissolve each pellet in 10 μl of water.
16. Add 1 μl of each size marker to 100 μl water.

3.3.2 Purification of RNA Fragments of Appropriate Length

Use siliconized microtubes while working with trace amounts of small RNAs to prevent the absorption of nucleic acids to the tube.

1. Add 10 μl of denaturing PAA gel loading solution to the RNA recovered in the Subheading 3.2.5, **step 6** as well as to 10 μl of 1:1 combined size marker described under Subheading 3.3.1, **step 16**.
2. Denature sample for 1 min at 90 °C.
3. Load the sample in a well on a 15% denaturing acrylamide gel with 7.5 M urea. Load size markers on opposite ends of the gel, framing the PAR-CLIP samples. Make sure to space different samples appropriately, typically at a two-well distance, to avoid cross-contamination.
4. Run the gel for 35 min at 30 W using 1× TBE buffer.
5. Dismantle the gel, leaving it mounted on one glass plate. Implant three tiny radioactive gel pieces asymmetrically at three of the four corners of the gel, to facilitate the alignment of the gel to the phosphorimager paper printout. Wrap the gel in plastic film (e.g., Saran wrap) to avoid contamination.

6. Expose the gel for at least 1 h to a phosphorimager screen.
7. Align the gel on top of a printout scaled to 100% according to the position of the three radioactive gel pieces. Excise the RNA between 19 and 35 nt. The size markers do not need to be excised.
8. Elute the RNA from the gel slices for 1 h at 37 °C with constant agitation with at least 400 μl of 0.3 M NaCl. Repeat the elution step.
9. Take off the supernatant, add 1 μl of glycogen solution and precipitate the small RNAs by adding 3 volumes of ethanol to the collected supernatant.
10. Precipitate the ligation products by incubation in a −20 °C freezer for 30 min and collect the pellet by centrifugation in a tabletop centrifuge at 4 °C at maximum speed (approx. 14,000 × *g*) for 15 min.
11. Discard the supernatant. Wash with 75% ethanol and centrifuge at 4 °C at maximum speed (approx. 14,000 × *g*) for 10 min.
12. Discard the supernatant. Collect residual ethanol by centrifugation at 14,000 × *g* for 1 min and discard the supernatant. Air-dry the RNA pellet for 5 min.
13. Dissolve the pellet in 10 μl water.

3.3.3 3′ Adapter Ligation

1. Prepare the following reaction mixture for ligation of the 3′ adenylated adapter, multiplying the volumes by the number of ligation reactions to be performed plus one extra volume to account for pipetting error:
   - 2 μl of 10× RNA ligase buffer without ATP
   - 6 μl 50% DMSO
   - 1 μl of 100 μM adenylated 3'adapter

   The number of ligation reactions is the number of samples plus two for the combined size markers (*see* Subheading 3.3.1, **step 16**), which are control for successful ligation and indicate the size range of the ligated RNA fragments that need to be recovered from the gel. Use 10 μl of combined size markers for each reaction.
2. Add 9 μl of the reaction mixture to each sample described under Subheading 3.3.2, **step 13**.
3. Denature the RNA by incubating the microtubes for 1 min at 90 °C. Place the microtube immediately on ice for 2 min.
4. Add 1 μl of Rnl2 (1–249) K227Q ligase (1 μg/μl), mix gently and incubate overnight on ice in the cold room.
5. Add 20 μl of denaturing PAA gel loading solution.

6. Denature sample for 1 min at 90 °C.
7. Load samples in two wells on a 15% denaturing acrylamide gel. Load unligated markers and the marker ligation reactions on each end of the gel, framing the PAR-CLIP samples. Space samples appropriately, typically at a two-well distance, to avoid cross-contamination.
8. Run the gel for 45 min at 30 W using 1× TBE buffer until the Bromophenol blue band is close to the bottom of the gel.
9. Dismantle the gel, leaving it mounted on one glass plate. Implant three tiny radioactive gel pieces asymmetrically at three of the four corners of the gel, to facilitate the alignment of the gel to the phosphorimager paper printout. Wrap the gel in plastic film (e.g., Saran wrap) to avoid contamination.
10. Expose the gel for at least 1 h to a phosphorimager screen.
11. Align the gel on top of a printout scaled to 100% according to the position of the three radioactive gel pieces. Cut out the RNA fragments ligated to the 3′ adapter using the ligated size markers as size range. Cut out the ligated size markers as well and proceed together with the other samples.
12. Elute the ligation product from the gel slices for 1 h at 37 °C with constant agitation with at least 400 μl of 0.3 M NaCl. Repeat the elution step.
13. Take off the supernatant, add 1 μl of glycogen solution and precipitate the small RNAs by adding three volumes of ethanol to the collected supernatant.
14. Allow the ligation products to precipitate by incubation in the freezer for 30 min and collect the pellet by centrifugation in a tabletop centrifuge at 4 °C at maximum speed (approx. 14,000 × *g*) for 15 min.
15. Discard the supernatant. Wash with microtube walls and pellet with 1 ml 75% ethanol and centrifuge at 4 °C at maximum speed (approx. 14,000 × *g*) for 10 min.
16. Discard the supernatant. Collect residual ethanol by centrifugation at 14,000 × *g* for 1 min and discard the supernatant. Air-dry the RNA pellet for 5 min.
17. Dissolve the pellet in 9 μl water.

3.3.4 5′ Adapter Ligation

1. Prepare the following reaction mixture, multiplying the volumes by the number of ligation reactions to be performed plus one extra volume to account for pipetting error:
   - 1 μl of 100 μM 5′ adapter oligoribonucleotide.
   - 2 μl of 10× RNA ligase buffer with ATP.
   - 6 μl 50% aqueous DMSO.

Add the 9 μl of this mixture to the sample described under Subheading 3.3.3, **step 17**. Prepare a ligation for only one of the ligated markers sample. Keep the other one to control the success of the ligation.

2. Denature the RNA by incubation for 1 min at 90 °C. Place the microtube immediately on ice for 2 min.
3. Add 2 μl of Rnl1, mix gently, and incubate for 1 h at 37 °C. Repeat this step twice.
4. Add 20 μl of denaturing PAA gel loading solution and incubate the samples for 2 min at 95 °C. Load the samples in two adjacent wells of a 20-well 12% acrylamide gel. As controls, load the unligated, 3′ adapter ligated and 3′ and 5′ adapter ligated size markers on the ends of the gel, framing the PAR-CLIP samples. Make sure to space different samples appropriately, typically at a two-well distance, to avoid cross-contamination.
5. Run the gel for 45 min at 30 W using 1× TBE buffer until the Bromophenol blue band is close to the bottom of the gel. Image the gel as described above and excise the new ligation product.
6. Elute the ligation product from the gel slices with 400 μl of 0.3 M NaCl by constant shaking at 1000 rpm for 1 h at 37 °C. Repeat the elution step.
7. Take off the supernatant, add 1 μl of glycogen solution and precipitate the small RNAs by adding 3 volumes of ethanol to the collected supernatant.
8. Allow the ligation products to precipitate by incubation in the freezer for 30 min and collect the pellet by centrifugation in a tabletop centrifuge at 4 °C at maximum speed (approx. 14,000 × *g*) for 15 min.
9. Discard the supernatant. Wash the microtube walls and pellet with 1 ml 75% ethanol and centrifuge at 4 °C at maximum speed (approx. 14,000 × *g*) for 10 min.
10. Discard the supernatant. Collect residual ethanol by centrifugation at 14,000 × *g* for 1 min and discard the supernatant. Air-dry the RNA pellet for 5 min.
11. Dissolve the pellet in 4.6 μl water.

3.3.5 Reverse Transcription

1. Prepare the following reaction mix, multiplying the volumes by the number of ligation reactions to be performed plus one extra volume to account for pipetting error:
   - 1 μl of 100 μM 3′ primer.
   - 1.5 μl 0.1 M DTT.
   - 3 μl 5× first-strand synthesis buffer.
   - 4.2 μl 10× dNTPs.

2. Denature the RNA described under Subheading 3.3.4, **step 11** by incubating the microtube for 30 s at 90 °C and transfer the microtube to a 50 °C incubator.
3. Add 8.7 μl of the prepared reaction mix to each sample and incubate for 3 min at 50 °C. Add 0.75 μl of SuperScript III reverse transcriptase and incubate for 2 h at 50 °C.
4. To hydrolyze the RNA, add 40 μl of 150 mM KOH/20 mM Tris base and incubate at 90 °C for 10 min.
5. Neutralize the solution by adding 40 μl of 150 mM HCl and check the pH of the mixture by spotting 1 μl on pH paper. The pH should be between 7.0 and 9.5 to ensure that the subsequent PCR is not inhibited. If necessary, readjust the pH by adding more base or acid.

3.3.6 PCR Amplification

1. Prepare the following mix multiplied with number of samples:
   - 40 μl of the 10× PCR buffer.
   - 40 μl 10× dNTPs.
   - 2 μl of 100 μM 5′ primer.
   - 2 μl of 100 μM 3′ primer.
   - 272 μl H_2O.

   Perform a standard 100 μl PCR with Taq polymerase (5 U/μl). 89 μl of the reaction mix will be used in a pilot PCR reaction and rest of the mixture will be used for the large scale PCR.
2. To 89 μl of the reaction mix add 10 μl from the cDNA solution described under Subheading 3.3.5, **step 6** and 1 μl of the Taq polymerase. Use the following cycle conditions:
   - 45 s at 94 °C.
   - 85 s at 50 °C.
   - 60 s at 72 °C.
3. To determine the necessary number of cycles for amplifying of the cDNA library, remove 10 μl aliquots every third cycle at the end of the 72 °C step starting with cycle number 15 up to cycle number 30.
4. Analyze the samples on a 2.5% agarose gel. The PCR-product might appear as a double band with the higher band running at the expected length of about 95–110 nt and a lower band at 65 nt corresponding to the direct ligation/template switch products of the 3′ adapter and 5′ adapter. Define the optimal cycle number for cDNA amplification, which should be within the exponential amplification phase of the PCR, approx. 3 cycles away from reaching the saturation level of PCR amplification.

5. Perform a 300 μl PCR with the optimal cycle number and analyze the product again on a 2.5% agarose gel.
6. Concentrate the DNA by ethanol precipitation and resuspend the pellet in 60 μl of 1× DNA loading dye.
7. Run the sample on two wells of a 2.5% agarose gel. Do not overload the gel to not compromise the separating capacity of the agarose gel.
8. Visualize the DNA on a UV transilluminator and excise the band corresponding to 95–110 nt with a clean scalpel.
9. Purify the DNA using the gel extraction kit according to the instructions of the manufacturer. Use 30 μl elution buffer to recover the cDNA.
10. Check the nucleic acid quality and quantity at an Agilent 2200 TapeStation with a High Sensitivity D1000 ScreenTape system. The size range of the sample should be around 95–110 nt in length.
11. Submit the amplified and purified cDNA for Illumina sequencing. The minimum amount is 10 μl of a 2 nM cDNA solution. We recommend using 50 bp sequencing on a HiSeq 2500 machine. Our adapters are compatible with the Illumina small RNA sequencing kit.

3.4 Computational Analyses

PAR-CLIP Suite v1.0 (https://rnaworld.rockefeller.edu/PARCLIP_suite/) is a quality control pipeline for PAR-CLIP datasets [23]. Using raw Illumina sequencing data, it identifies the predominant target RNA category or categories for the RBP of interest and provides the T-to-C conversion frequency resolved by read length and RNA category. The read annotation is performed in a hierarchical manner, where reads mapping to more than one category are assigned to a single category according to cellular abundance of the category members unless the annotation hierarchy is modified by the user. Figure 1 depicts the histogram output for a PAR-CLIP experiment studying IGF2BP2 protein [12]; Fig. 1a provides an overview of read composition resolved by read length allowing for up to two mismatches in mapping, while Fig. 1b–e further resolved the mapped reads of the predominant target RNA categories by error distance 0 (d0), error distance 1 (d1; split in T-to-C and d1 other than T-to-C), and error distance 2 (d2). This process readily discriminates true target RNA categories from non-cross-linked background RNA categories populated by fragments of abundant cellular RNAs.

The abundance of cross-linked reads over all d1 reads per RNA categories may be used for unsupervised clustering of RBPs previously studied by PAR-CLIP as shown in Table 1 and Fig. 2. RBPs sharing the same predominant target RNA categories cluster together. T-to-C conversion ratios and number of cross-linked

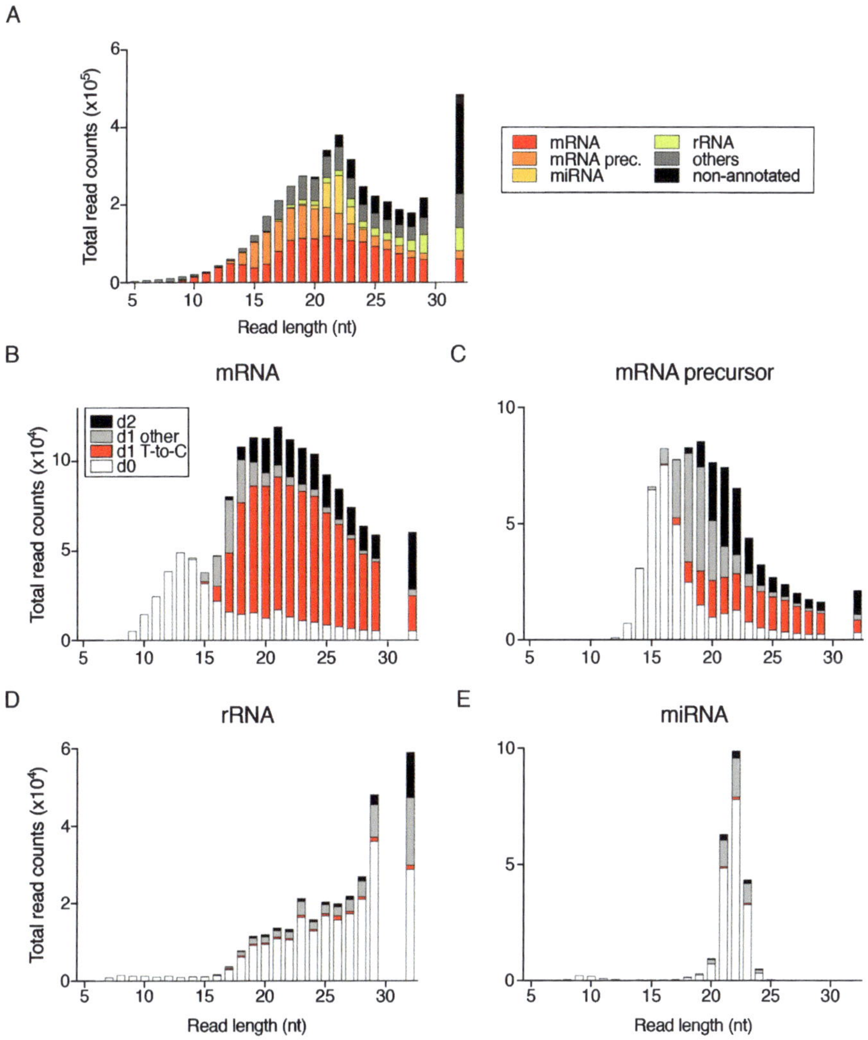

Fig. 1 (**a**) Composition of IGF2BP2 (SRR048957) PAR-CLIP sequence reads by RNA categories mapping to each RNA category with up to two mismatches represented by *stacked bar graphs* and resolved by length of adapter-extracted sequence reads. Composition of the most abundant RNA categories assigned as d0 (*white*), d1 T-to-C (*red*), d1 other than T-to-C, (*light grey*), and d2 (*black*) for mRNA (**b**), mRNA prec. (**c**), rRNA (**d**), and miRNA (**e**), respectively

reads may also be displayed to summarize the results of the various experiments.

Once the predominant category of targeted RNA(s) is identified, additional computation may be conducted to identify binding sites and underlying RNA recognition motif(s) of each RBP. Programs such as PARalyzer [24], CLIPZ [25], starBase [26], WavclusteR [27], doRina [28], Piranha [29], miRTarCLIP [30], PIPE-CLIP [31], dCLIP [32], and PARA-suite [33] have been developed for this purpose. Binding sites and RNA recognition motifs may be further validated by biochemical studies with

Table 1
PAR-CLIP libraries used for the hierarchical clustering analysis

#	Dataset name	Accession number	Reference
01	AGO2[1]	SRR189786	[36]
02	AGO2[2]	SRR189787	[36]
03	ATXN2	DRR014103	[37]
04	CAPRIN1[1]	SRR500484	[38]
05	CAPRIN1[2]	SRR500485	[38]
06	CPSF1	SRR488737	[39]
07	CSTF2	SRR488740	[39]
08	CSTF2T	SRR488741	[39]
09	ELAVL1	SRR653239	[40]
10	FBL[1]	SRR651725	[41]
11	FBL[2]	SRR651726	[41]
12	FMR1	SRR527727	[13]
13	FUS[1]	SRR070449	[34]
14	FUS[2]	SRR070450	[34]
15	FXR1	SRR527731	[13]
16	IGF2BP2	SRR048957	[12]
17	IGF2BP3[1]	SRR048962	[12]
18	IGF2BP3[2]	SRR048963	[12]
19	MOV10[1]	SRR921441	[42]
20	MOV10[2]	SRR921442	[42]
21	NOP56	SRR651724	[41]
22	NOP58[1]	SRR651722	[41]
23	NOP58[2]	SRR651723	[41]
24	PUM2[1]	SRR048967	[12]
25	PUM2[2]	SRR048968	[12]
26	RBPMS	SRR1188147	[43]
27	RC3H1[1]	SRR857933	[44]
28	RC3H1[2]	SRR857934	[44]
29	SSB	SRR4301753	[23]

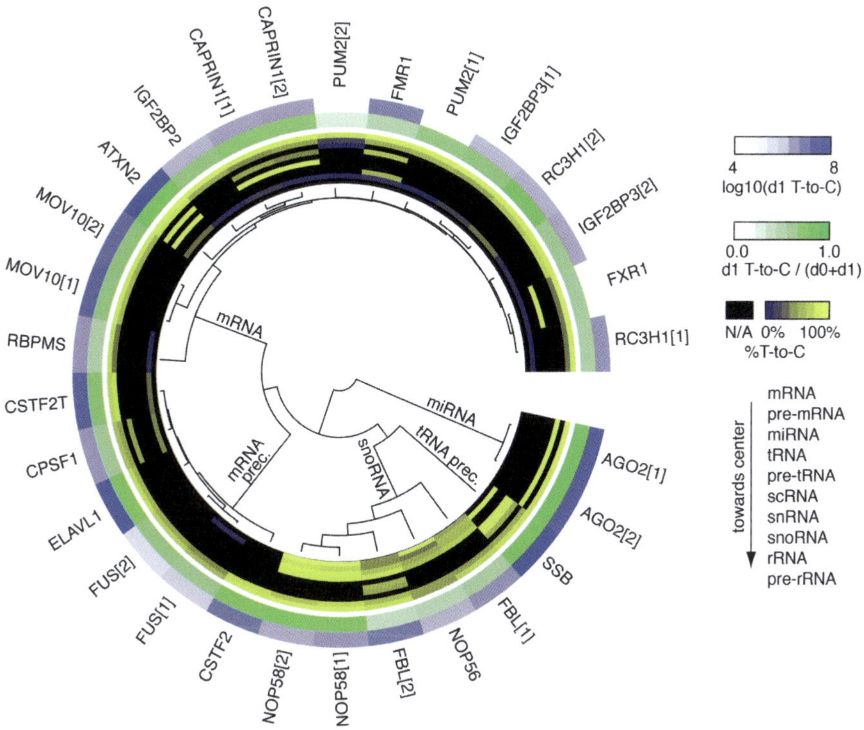

Fig. 2 Comparison of PAR-CLIP libraries of various RBPs organized by unsupervised clustering of the relative abundance of cross-linked reads per RNA category, whereby the initial distances between each pair of libraries were calculated as Euclidian distance between vectors of fractions of reads with T-to-C conversion over all d1 reads $\geq$20 nt for each RNA category. RBPs with similar target spectra cluster together and individual differences are indicated by the branch lengths of the tree-like diagram placed in the *center of the circle*. The predominant cross-linked RNA categories are indicated on the major branches. The *blue* and *yellow circular rings* report the ratio of d1 T-to-C reads over all d1 reads for each RNA category, unless the d1 T-to-C reads in the next lower count category are less than 0.75% of the major cross-linked RNA category, which are then indicated in *black*. The *green color scheme* reports the overall ratio of d1 T-to-C reads over all (d0 + d1) reads providing information about non-cross-linked background, whereas the *blue color scheme* of the *outer circle* indicates the total number of d1 T-to-C reads per library

recombinant RBP [13, 34]. Finally RNA targets may be binned and ranked according to number of binding sites or normalized cross-linked read counts and correlations of regulatory effects on target RNA stabilization as a function of overexpression or knock-down/out may be assessed to provide functional insights [13].

4 Notes

1. We recommend performing pilot PAR-CLIP experiments before performing the large-scale experiment. The experimentalist will use the pilot experiment to optimize RNase and IP

steps to obtain the conditions which RNA fragments and IP efficacy are optimum.

2. To determine the cross-linking efficiency for a given RBP, we recommend using a previously characterized RBP as positive control. Plasmids encoding FLAG/HA-tagged RBPs are available at www.addgene.org.
3. 4SU incorporation rate into cellular RNA is cell-type-dependent. The concentration of photoreactive nucleoside supplementing the cell culture medium and the culture incubation time needs to be optimized to arrive at incorporation rates ≥1% relative to uridine.
4. The incorporation efficiency of 4SU into RNAs is quantified using high-performance liquid chromatography (HPLC) [35] or quantitative mass spectrometry [20] after enzymatic digestion of total RNA to ribonucleosides.
5. Alternatively, other photoactivatable ribonucleosides such 6-thioguanosine (6SG) can be used. 6SG has a lower cross-linking efficiency in nascent RNAs compared to 4SU and it is more toxic. Additionally, the computational pipeline needs to be adjusted to accommodate G-to-A conversion at the cross-linking site [12].
6. For cells grown in suspension, they must be collected and resuspended in a small volume of ice-cold PBS and transferred into a tissue culture dish before they are irradiated at 365 nm.
7. The RNA footprint size depends on the type of RNase used as well as the selected RBP. RNA concentration in the cell lysate, the type of RNase and concentration, and the incubation time and temperature influence the degree of RNA fragmentation. Consequently, downstream steps such as RBP IP efficiency, the preparation of cDNA libraries, sequencing results, and data analyses, are impacted. It is therefore necessary to optimize the RNase treatment conditions for each family of RBP. Final RNase concentrations typically vary over a range of 0.1–1 U/μl in the lysate.
8. RNase treatment should be optimized to obtain RNA fragments of 20–40 nt in length. Most reads longer than 20 nt can be mapped uniquely to reference RNAs, while shorter reads do not, making localization of cross-linked positions impossible.
9. For experiments with less than 2.5 ml of cell lysate do not use less than 25 μl of magnetic beads or bead recovery during the experiments is compromised resulting in lower than expected yields.
10. For FLAG tagged proteins conjugated ANTI-FLAG M2 Magnetic Beads (Sigma) may be used.

11. High IP efficacies are critical to recover sufficient cross-linked RNA–RBP complexes. The IP efficacy is highly dependent on the specific antibody as well as the method used for its matrix-coupling. We recommend validating the cross-linking efficacy of the RBP first using FLAG/HA-tagged cell lines. In the Methods section we use anti-FLAG M2 antibody, but if specific antibodies to the RBP of interest are available, these may be used to IP endogenous RBP in different cell lines.
12. The supernatant after IP can be used to immunoprecipitate different RBP using specific antibodies if those RBP were not depleted from the lysate in the first IP process.
13. To prevent oxidation of 4SU during RNA isolation and library preparation, washing buffers require the addition of 0.1 mM dithiothreitol (DTT) before use. However, DTT might also damage the magnetic beads. Therefore the duration of wash steps should not be extended.
14. Antibodies have a maximal tolerated salt concentration for antigen binding. This critical salt concentration must be determined before starting PAR-CLIP and should not be exceeded during and after IP. We recommend performing a pilot IP and washing the matrix with different salt concentrations to detect any elution of the RBP during the washing procedure by western blotting.
15. The second RNase treatment described under Subheading 3.2.2 is performed after IP and it is optional. This step might be necessary to minimize the size of RNA fragments, but it might also result in excessive fragmentation obtaining RNA fragments shorter than the optimum 20 nt.
16. The dephosphorylation buffer used under Subheading 3.2.2 contains the same composition than the commercial 1× NEB buffer 3.
17. If there is evidence of occlusion or modification of the 5′ end of bound RNAs, alternative radiolabeling methods can be used, such as labeling the 3′ end of RNA using T4 RNA ligase I and [^{32}p]Cp. To facilitate conversion of 5′ modified RNA into cDNA using standard adapter ligation protocol described here, both 5′ and 3′ termini of recovered RNA need to be converted to comprise 5′ phosphate and 3′ hydroxyl.
18. The cross-linked RNA–protein complexes might be recovered directly from the SDS gel by electroelution, or alternatively, the RNA–protein complexes can be transferred to a nitrocellulose membrane and subsequently recovered from the membrane by performing the Proteinase K digestion directly on the excised piece of membrane that corresponds to the expected size of the RBP.

References

1. Martin KC, Ephrussi A (2009) mRNA localization: gene expression in the spatial dimension. Cell 136:719–730. https://doi.org/10.1016/j.cell.2009.01.044
2. Moore MJ, Proudfoot NJ (2009) Pre-mRNA processing reaches back to transcription and ahead to translation. Cell 136:688–700. https://doi.org/10.1016/j.cell.2009.02.001
3. Sonenberg N, Hinnebusch AG (2009) Regulation of translation initiation in eukaryotes: mechanisms and biological targets. Cell 136:731–745. https://doi.org/10.1016/j.cell.2009.01.042
4. Gerstberger S, Hafner M, Tuschl T (2014) A census of human RNA-binding proteins. Nat Rev Genet 15:829–845. https://doi.org/10.1038/nrg3813
5. Greenberg JR (1979) Ultraviolet light-induced crosslinking of mRNA to proteins. Nucleic Acids Res 6:715–732
6. Wagenmakers AJ, Reinders RJ, van Venrooij WJ (1980) Cross-linking of mRNA to proteins by irradiation of intact cells with ultraviolet light. Eur J Biochem 112:323–330
7. Mayrand S, Setyono B, Greenberg JR, Pederson T (1981) Structure of nuclear ribonucleoprotein: identification of proteins in contact with poly(A)+ heterogeneous nuclear RNA in living HeLa cells. J Cell Biol 90:380–384
8. Dreyfuss G, Choi YD, Adam SA (1984) Characterization of heterogeneous nuclear RNA-protein complexes in vivo with monoclonal antibodies. Mol Cell Biol 4:1104–1114
9. Konig J, Zarnack K, Luscombe NM, Ule J (2011) Protein-RNA interactions: new genomic technologies and perspectives. Nat Rev Genet 13:77–83. https://doi.org/10.1038/nrg3141
10. Gerstberger S, Hafner M, Tuschl T (2013) Learning the language of post-transcriptional gene regulation. Genome Biol 14:130. https://doi.org/10.1186/gb-2013-14-8-130
11. Mann M (2006) Functional and quantitative proteomics using SILAC. Nat Rev Mol Cell Biol 7:952–958. https://doi.org/10.1038/nrm2067
12. Hafner M, Landthaler M, Burger L et al (2010) Transcriptome-wide identification of RNA-binding protein and microRNA target sites by PAR-CLIP. Cell 141:129–141. https://doi.org/10.1016/j.cell.2010.03.009
13. Ascano M, Mukherjee N, Bandaru P et al (2012) FMRP targets distinct mRNA sequence elements to regulate protein expression. Nature 492:382–386. https://doi.org/10.1038/nature11737
14. Flores O, Nakayama S, Whisnant AW et al (2013) Mutational inactivation of herpes simplex virus 1 microRNAs identifies viral mRNA targets and reveals phenotypic effects in culture. J Virol 87:6589–6603. https://doi.org/10.1128/JVI.00504-13
15. Wagschal A, Najafi-Shoushtari SH, Wang L et al (2015) Genome-wide identification of microRNAs regulating cholesterol and triglyceride homeostasis. Nat Med 21:1290–1297. https://doi.org/10.1038/nm.3980
16. Llobet-Navas D, Rodríguez-Barrueco R, La Iglesia-Vicente de J et al (2014) The microRNA 424/503 cluster reduces CDC25A expression during cell cycle arrest imposed by transforming growth factor β in mammary epithelial cells. Mol Cell Biol 34:4216–4231. https://doi.org/10.1128/MCB.00611-14
17. Hansen HT, Rasmussen SH, Adolph SK et al (2015) Drosophila Imp iCLIP identifies an RNA assemblage coordinating F-actin formation. Genome Biol 16:123. https://doi.org/10.1186/s13059-015-0687-0
18. Xiong X-P, Vogler G, Kurthkoti K et al (2015) SmD1 modulates the miRNA pathway independently of its pre-mRNA splicing function. PLoS Genet 11:e1005475. https://doi.org/10.1371/journal.pgen.1005475
19. Lebedeva S, Jens M, Theil K et al (2011) Transcriptome-wide analysis of regulatory interactions of the RNA-binding protein HuR. Mol Cell 43:340–352. https://doi.org/10.1016/j.molcel.2011.06.008
20. Jungkamp A-C, Stoeckius M, Mecenas D et al (2011) In vivo and transcriptome-wide identification of RNA binding protein target sites. Mol Cell 44:828–840. https://doi.org/10.1016/j.molcel.2011.11.009
21. Creamer TJ, Darby MM, Jamonnak N et al (2011) Transcriptome-wide binding sites for components of the Saccharomyces cerevisiae non-poly(A) termination pathway: Nrd1, Nab3, and Sen1. PLoS Genet 7:e1002329. https://doi.org/10.1371/journal.pgen.1002329
22. Baejen C, Torkler P, Gressel S et al (2014) Transcriptome maps of mRNP biogenesis factors define pre-mRNA recognition. Mol Cell 55:745–757. https://doi.org/10.1016/j.molcel.2014.08.005
23. Garzia A, Meyer C, Morozov P et al (2016) Optimization of PAR-CLIP for transcriptome-wide identification of binding sites of RNA-binding proteins. Methods 118–119:24–40. https://doi.org/10.1016/j.ymeth.2016.10.007

24. Corcoran DL, Georgiev S, Mukherjee N et al (2011) PARalyzer: definition of RNA binding sites from PAR-CLIP short-read sequence data. Genome Biol 12:R79. https://doi.org/10.1186/gb-2011-12-8-r79
25. Khorshid M, Rodak C, Zavolan M (2011) CLIPZ: a database and analysis environment for experimentally determined binding sites of RNA-binding proteins. Nucleic Acids Res 39: D245–D252. https://doi.org/10.1093/nar/gkq940
26. Yang J-H, Li J-H, Shao P et al (2011) starBase: a database for exploring microRNA-mRNA interaction maps from argonaute CLIP-Seq and degradome-Seq data. Nucleic Acids Res 39:D202–D209. https://doi.org/10.1093/nar/gkq1056
27. Sievers C, Schlumpf T, Sawarkar R et al (2012) Mixture models and wavelet transforms reveal high confidence RNA-protein interaction sites in MOV10 PAR-CLIP data. Nucleic Acids Res 40:e160–e160. https://doi.org/10.1093/nar/gks697
28. Anders G, Mackowiak SD, Jens M et al (2012) doRiNA: a database of RNA interactions in post-transcriptional regulation. Nucleic Acids Res 40:D180–D186. https://doi.org/10.1093/nar/gkr1007
29. Uren PJ, Bahrami-Samani E, Burns SC et al (2012) Site identification in high-throughput RNA-protein interaction data. Bioinformatics 28:3013–3020. https://doi.org/10.1093/bioinformatics/bts569
30. Chou C-H, Lin F-M, Chou M-T et al (2013) A computational approach for identifying microRNA-target interactions using high-throughput CLIP and PAR-CLIP sequencing. BMC Genomics 14(Suppl 1):S2. https://doi.org/10.1186/1471-2164-14-S1-S2
31. Chen B, Yun J, Kim MS et al (2014) PIPE-CLIP: a comprehensive online tool for CLIP-seq data analysis. Genome Biol 15:R18. https://doi.org/10.1186/gb-2014-15-1-r18
32. Wang T, Xie Y, Xiao G (2014) dCLIP: a computational approach for comparative CLIP-seq analyses. Genome Biol 15:R11. https://doi.org/10.1186/gb-2014-15-1-r11
33. Kloetgen A, Borkhardt A, Hoell JI, McHardy AC (2016) The PARA-suite: PAR-CLIP specific sequence read simulation and processing. Peer J 4:e2619–e2622. https://doi.org/10.7717/peerj.2619
34. Hoell JI, Larsson E, Larsson E et al (2011) RNA targets of wild-type and mutant FET family proteins. Nat Publ Group 18:1428–1431. https://doi.org/10.1038/nsmb.2163
35. Spitzer J, Hafner M, Landthaler M et al (2014) PAR-CLIP (photoactivatable ribonucleoside-enhanced crosslinking and immunoprecipitation): a step-by-step protocol to the transcriptome-wide identification of binding sites of RNA-binding proteins. Meth Enzymol 539:113–161. https://doi.org/10.1016/B978-0-12-420120-0.00008-6
36. Kishore S, Jaskiewicz L, Burger L et al (2011) A quantitative analysis of CLIP methods for identifying binding sites of RNA-binding proteins. Nat Meth 8:559–564. https://doi.org/10.1038/nmeth.1608
37. Yokoshi M, Li Q, Yamamoto M et al (2014) Direct binding of ataxin-2 to distinct elements in 3' UTRs promotes mRNA stability and protein expression. Mol Cell 55:186–198. https://doi.org/10.1016/j.molcel.2014.05.022
38. Baltz AG, Munschauer M, Schwanhäusser B et al (2012) The mRNA-bound proteome and its global occupancy profile on protein-coding transcripts. Mol Cell 46:674–690. https://doi.org/10.1016/j.molcel.2012.05.021
39. Martin G, Gruber AR, Keller W, Zavolan M (2012) Genome-wide analysis of pre-mRNA 3″ end processing reveals a decisive role of human cleavage factor I in the regulation of 3″ UTR length. Cell Rep 1:753–763. https://doi.org/10.1016/j.celrep.2012.05.003
40. Mandal PK, Ewing AD, Hancks DC, Kazazian HH (2013) Enrichment of processed pseudogene transcripts in L1-ribonucleoprotein particles. Hum Mol Genet 22:3730–3748. https://doi.org/10.1093/hmg/ddt225
41. Kishore S, Gruber AR, Jedlinski DJ et al (2013) Insights into snoRNA biogenesis and processing from PAR-CLIP of snoRNA core proteins and small RNA sequencing. Genome Biol 14: R45. https://doi.org/10.1186/gb-2013-14-5-r45
42. Gregersen LH, Schueler M, Munschauer M et al (2014) MOV10 Is a 5′ to 3′ RNA helicase contributing to UPF1 mRNA target degradation by translocation along 3′ UTRs. Mol Cell 54:573–585. https://doi.org/10.1016/j.molcel.2014.03.017
43. Farazi TA, Leonhardt CS, Mukherjee N et al (2014) Identification of the RNA recognition element of the RBPMS family of RNA-binding proteins and their transcriptome-wide mRNA targets. RNA 20:1090–1102. https://doi.org/10.1261/rna.045005.114
44. Murakawa Y, Hinz M, Mothes J et al (2015) RC3H1 post-transcriptionally regulates A20 mRNA and modulates the activity of the IKK/NF-κB pathway. Nat Commun 6:7367. https://doi.org/10.1038/ncomms8367

Chapter 6

Characterizing mRNA Sequence Motifs in the 3′-UTR Using GFP Reporter Constructs

Rene Geissler and Andrew Grimson

Abstract

GFP reporter constructs are widely used as an expression system for studying the function of regulatory sequence motifs (*cis* elements) within the 3′-UTRs (3′ untranslated regions) of mRNAs. Here we provide details on the characterization of individual sequence motifs, which typically regulate mRNA decay and translation. In addition, we describe methods to identify *trans* factors required for the function of such elements. To facilitate efficient identification of novel functional 3′-UTR motifs, we describe a screening approach based on dual-color fluorescence reporter constructs. Such screening approaches can be used to test large collections of defined sequence or libraries of random sequences.

Key words Posttranscriptional gene regulation, 5′ and 3′ untranslated regions, *cis*-regulatory elements, Dual-color high-throughput assay, mRNA decay, Translation, Functional genomics, RNA-seq, shRNAs, Lentivirus

1 Introduction

Posttranscriptional control of gene expression plays essential roles in development, metabolism, the pathogenesis of diseases, including many cancers, and evolution. Posttranscriptional regulation is mediated, primarily, by altering the mRNA decay and translation rates [1]. While translation rates are often regulated by regulatory elements within the 5′ untranslated region (UTR) of mRNAs, decay rates are predominantly modulated by elements located in the 3′-UTRs [2]. In particular, 3′-UTRs are a genomic hotspot for regulatory motifs [3, 4], which recruit *trans*-acting factors (e.g., RNA-binding proteins or miRNAs) to create a complex collection of posttranscriptional events that act on different stages of mRNA metabolism, including maturation, nuclear export, stability, and translation [5–11]. The majority of well-characterized 3′-UTR motifs, e.g., miRNA target sites and AU-rich elements, are destabilizing motifs and function by recruiting RNA decay enzymes to the 3′ end of mRNAs [12–22]. In mammals, mRNA decay is

Shireen R. Lamandé (ed.), *mRNA Decay: Methods and Protocols*, Methods in Molecular Biology, vol. 1720,
https://doi.org/10.1007/978-1-4939-7540-2_6,

preferentially initiated by poly(A) tail shortening mediated by specific deadenylases, including the major deadenylase complexes CCR4-NOT, PAN2-PAN3 and PARN [2]. Following deadenylation, mRNAs are further degraded by removing the 5′ cap structure through the decapping complex DCP1/DCP2. Subsequently, recruitment of the exonuclease XRN1 leads to degradation of the mRNA in a 5′–3′ direction. In comparison to the 5′–3′ mRNA decay pathway, motifs which promote 3′–5′ decay after deadenylation recruit the exosome or exonucleases from the DIS3/3L protein family [23–28].

Efficient identification and characterization of motifs involved in translation and mRNA decay has been facilitated by the generation of high-throughput sequencing platforms, which permit the analyses of sequence motifs in large scale. Together with dual-color fluorescence-based reporter constructs, thousands of functional sequence motifs can be investigated simultaneously. The dual-color reporter system, e.g., dsRed/GFP-3′-UTR constructs, is stably integrated into the genome at a specific locus. This approach enables the direct comparison of the regulatory potential of each motif with a pool of motifs, since each construct is expressed within the same genomic background. These approaches have advanced the pace of discovery in posttranscriptional biology research, and have been used widely to characterize functional sequence elements [29–31].

Here, we describe procedures for characterizing individual sequence motifs located in 3′-UTRs that regulate mRNA decay and translation. In addition, we provide details to analyze the decay pathways in which destabilizing motifs are involved. We further describe the generation of a GFP-3′-UTR reporter library containing randomized sequence motifs in 3′-UTRs. We focus on a method that controls the site of integration into the genome, thereby limiting reporter transcriptional variation attributed to genomic location, and which is appropriate for generating a population of many thousands of individually integrated reporters in parallel.

2 Materials

2.1 General

1. Cell lines: Flp-In™ T-REx™ cells (ThermoFisher Scientific), Hek293T, human cell lines of interests (e.g., Huh7, A549, MCF7).
2. Plasmids: Derived from pEF5/FRT/V5-D-TOPO® or pcDNA5/FRT (ThermoFisher Scientific), and containing a dsRed/GFP-3′-UTR expression cassette, used to generate cell lines stably expressing the reporter constructs (*see* **Note 1**), which are integrated at a specific gene locus when paired with Flp-In™

T-REx™ cells. pOG44 (ThermoFisher Scientific), for expression of the Flp recombinase. Lentiviral vectors for expressing individual GFP-3′-UTR mRNAs and shRNAs; packaging and envelop plasmids for generating lentiviral particles (*see* **Note 1**).

3. Antibiotics: hygromycin (50 mg/ml stock solution), zeocin (100 mg/ml stock solution), penicillin (10,000 Units/ml), streptomycin (10 mg/ml stock solution), geneticin (50 mg/ml stock solution), puromycin (10 mg/ml stock solution), ampicillin (100 mg/ml) (*see* **Note 2**).
4. PCR reagents: sterile deionized distilled water H_2O (H_2O_{dd}), oligonucleotides, template (Human DNA or cDNA), dNTPs, Phusion® High-Fidelity DNA Polymerase, Phusion® High-Fidelity DNA Polymerase buffer, qPCR reaction mix.
5. Enzymes: DNA Polymerase I—Large (Klenow) Fragment, Restriction enzymes (XhoI, BamHI), T4 DNA Ligase, Reverse Transcriptase.
6. Heat-shock ultracompetent *E. coli* cells.
7. Reagents for polyacrylamide gel purification (PAGE): 30% (w/v) acrylamide:bisacrylamide (29:1), ammonium persulfate (APS), tetramethylethylenediamine (TEMED), Tris–borate–EDTA buffer (TBE—5×: 54 g Tris base, 27.5 g boric acid, 20 ml of 0.5 M EDTA, pH 8.3, bring volume to 1 l with H_2O_{dd}).
8. Cell culture transfection media: e.g., Lipofectamine 2000 (ThermoFisher Scientific).
9. Cell strainer.
10. TRIzol (ThermoFisher Scientific) for RNA preparation.
11. DNA extraction kit (e.g., Puregene Core Kit A from Qiagen or GeneJET Genomic DNA Purification Kit from ThermoFisher Scientific).
12. Lentiviral detection and quantification kits (e.g., Lentivirus Rapid Quantitation Kit from Cellbiolabs or Lenti-X™ qRT-PCR Titration Kit from Takara).
13. Polysomal analysis: cycloheximide, gradient mixer (e.g., Gradient Master—BioComp), ultracentrifuge (e.g., Beckman Coulter—SW41 rotor), gradient fractionation system (e.g., BR-188—Brandl, or Density Gradient Fractionator—Isco).

2.2 Media

1. LB medium: Dissolve 10 g NaCl, 10 g Tryptone, 5 g yeast extract and adjust pH to 7.0 with NaOH. Bring volume to 1 l with H_2O_{dd}. Autoclave for 30 min.
2. Cell culture media: DMEM supplemented with 10% fetal bovine serum (FBS), 1% penicillin/streptomycin, and 100 μg/mL zeocin for Flp-In™ T-REx™ cells. After stable

integration of the pEF5/FRT/V5-D-TOPO® or pcDNA5/FRT plasmids, cells are maintained in media containing 80 μg/mL hygromycin, omitting zeocin. For generation of lentiviral particles, DMEM supplemented with 30% FBS and 1% penicillin/streptomycin is suggested.

2.3 Buffer

1. Phosphate-buffered saline (PBS): 10 mM PO_4^{3-}, 137 mM NaCl, and 2.7 mM KCl, pH 7.4.
2. Cell sorting solution: PBS supplemented with 2 mM EDTA and 1% FBS (*see* **Note 3**).
3. Oligonucleotide annealing buffer: 50 mM NaCl, 10 mM Tris–HCl, 10 mM $MgCl_2$, 1 mM DTT pH 7.9, identical to New England Biolabs buffer 2 (NEB 2).
4. Polysome buffer: 10 mM HEPES, 100 mM KCl, 5 mM $MgCl_2$, 100 μg/ml cycloheximide, 5 mM DTT, pH 7.6.
5. Cell lysis buffer for polysomal analysis: Polysome buffer supplemented with 1% Triton X-100.

3 Methods

3.1 Generation of GFP-3′-UTR and shRNA Lentiviral Particles

1. Seed 1.5×10^5 Hek293T cells in 6 cm tissue culture plates in a total volume of 5 ml of DMEM supplemented with 10% FBS.
2. After ~24 h, transfect cells with 0.9 μg packaging plasmid, 0.1 μg envelope plasmid, and 1 μg GFP-3′-UTR expression construct or shRNA plasmid (will be used later to identify decay factors that are recruited by the motifs—*see* Subheading 3.5) according to the manufacturer's instructions (*see* **Note 4**).
3. Incubate cells for 18–24 h.
4. Change media to DMEM supplemented with 30% FBS and 1% penicillin/streptomycin (*see* **Note 5**).
5. Harvest media containing lentiviral particles 30 h after media change to 30% FBS.
6. Transfer media to Falcon tube and spin at $400 \times g$ for 3 min.
7. Analyze the lentiviral titers using lentiviral titration kits.
8. Snap-freeze aliquots and store at −80 °C.

3.2 Generation of Stably Integrated Cells Expressing GFP-3′-UTR Constructs and shRNAs

1. Seed 10^5 cells of desired cell line in a well of a 6-well tissue culture plate in a total volume of 2 ml of DMEM supplemented with 10% FBS and 1% penicillin/streptomycin.
2. Remove media 24 h later and replace with DMEM supplemented with 10% FBS, 1% penicillin/streptomycin, and 8 μg/ml polybrene.
3. Add virus to cells of interests (*see* **Note 6**).

4. After 24 h, remove media and replace with DMEM supplemented with 10% FBS, 1% penicillin/streptomycin and the selecting antibiotics.
5. RNA levels of cell lines stably expressing GFP-3′-UTR constructs can be analyzed 3 days post lentiviral infection. When infecting with lentiviral particles containing shRNA constructs, test for knock down efficiency 3–5 days after lentiviral infection using Western blot or qPCR.

3.3 Analyzing Motifs Involved in Translation

1. Seed cells expressing GFP-3′-UTR constructs, and in parallel a control version in which the sequence motif of interest is mutated, into five 10 cm tissue culture plates so that they are 70–80% confluent the next day.
2. Remove media and incubate cells with DMEM supplemented with 10% FBS, 1% penicillin/streptomycin and 100 μg/ml cycloheximide for 3 min at growth temperature.
3. Remove media and wash cells with ice-cold PBS supplemented with 100 μg/ml cycloheximide and keep plates one ice.
4. Remove PBS (*see* **Note 7**).
5. Add 500 μl lysis buffer to one 10 cm tissue culture plate.
6. Scrape cells and transfer lysate to the next plate.
7. Continue so that all plates are lysed with the starting 500 μl lysis buffer.
8. After the final plate is scraped, transfer lysate into a 1.5 ml tube.
9. Centrifuge lysate at 12,000 × *g* for 10 min at 4 °C.
10. Transfer supernatant to a new 1.5 ml tube (*see* **Note 8**).
11. For polysome profiling, prepare 15% and 45% (w/v) sucrose solutions in polysome buffer.
12. Prepare gradients in gradient maker.
13. Load 500 μl of cell lysate onto the gradient.
14. Centrifuge at 247,000 × *g* for 100 min at 4 °C in an SW41 rotor.
15. Fractionate gradient at 0.75 ml/min with continual monitoring of OD_{254} values.
16. Aliquot and snap freeze gradient fractions or proceed to RNA isolation using 1 ml TRIzol and 100 μl from each gradient fraction.
17. Proceed to cDNA synthesis using oligo(dT)-primed RNA followed by quantitative PCR (qPCR).
18. Calculate mRNA levels in each gradient fraction using housekeeping genes (e.g., GAPDH, β-Actin) as a reference (*see* **Note 9**).

3.4 Analyzing Motifs Involved in mRNA Decay

1. For each time point (3–5, excluding time point 0 h), seed 5×10^5 cells into a well of a 6-well plate.
2. Replace media with DMEM supplemented with 10% FBS, 1% penicillin/streptomycin, and 2.5 μg/ml actinomycin D; return cells to incubator (*see* **Note 10**).
3. At each time point, remove media and wash cells once with PBS.
4. Lyse cells by adding 1 ml of TRIzol.
5. Proceed to RNA isolation, cDNA synthesis, and qPCR as outlined in **steps 17** and **18** in Subheading 3.3.
6. Calculate the mRNA half-lives ($t_{1/2}$) using the equation: $t_{1/2} = \log_e 2/k$. The decay rate constant, k, can be determined from the slope of a plot of mRNA level ($\log_2$) as a function of time (Chen et al. [32]).

3.5 Identification of the RNA Decay Pathway Accelerated by Destabilizing Motifs

1. Infect stably integrated cells expressing individual GFP-3′-UTR constructs with lentiviral particles containing shRNA expression constructs (subunits of the CCR4-NOT complex, PAN2-PAN3, exosome or decapping complex, PARN, XRN1, DIS3L, DIS3L2, control shRNAs) according to Subheading 3.2.
2. Harvest cells 4 d post infection using TRIzol.
3. Proceed to RNA isolation.
4. Perform cDNA synthesis using an oligo(dT) and/or gene-specific primers (*see* **Note 11**).
5. For individual GFP-3′-UTR constructs analyze RNA levels using qPCR (*see* **Note 11**).

3.6 Generation of a Randomized 3′-UTR Motif Library

1. Design oligonucleotides with the desired length of randomized sequences, containing an XhoI site, the 6–12 nucleotide random sequence, a BamHI site, a hairpin sequence, and a second BamHI site.

   5′ atatatCTCGAGatatN_{6-12}gtatGGATCCattagtaataatGGATCCatac 3′ (N = A, T, G, or C) (*see* **Note 12**).

2. For annealing the oligonucleotide, denature the DNA at 95 °C for 5 min in the presence of 50 mM NaCl, 10 mM Tris–HCl, 10 mM $MgCl_2$, 1 mM DTT pH 7.9 supplemented with 2 mM dNTPs.
3. Cool on ice for 3 min. This step generates a partially double-stranded oligonucleotide substrate for the DNA polymerase.
4. Fill in the 3′ recessed ends using DNA Polymerase I (Large Klenow Fragment) at 25 °C for 15 min.
5. PAGE-purify the double-stranded oligonucleotide on a 12% nondenaturing gel.

6. Digest the double-stranded oligonucleotide with XhoI and BamHI.
7. Repeat the PAGE purification.
8. Digest pEF5/FRT/V5-D-TOPO®-GFP-dsRED or pcDNA5/FRT-GFP-dsRED with XhoI and BamHI.
9. Gel extract the digested plasmid.
10. Ligate 50 ng of digested plasmid and 2.3 ng of digested oligonucleotide with T4 DNA Ligase at 16 °C for 6 h.
11. Transform ligation reaction into heat-shock ultracompetent *E. coli* cells.
12. Transfer cells to 250 ml LB-Ampicillin.
13. Prepare plasmids using a plasmid maxiprep system.
14. Validate plasmid libraries using Illumina sequencing. Amplify a region spanning the 3′ end of the GFP-coding sequence and 5′ end of the Zeocin resistance gene using Phusion® High-Fidelity DNA Polymerase.
15. Reamplify the PCR product (region surrounding the variable nucleotide site) to add individual barcodes and Illumina sequencing adaptors.
16. Sequence the libraries to generate 50 nt reads.

3.7 Stable Integration of GFP-3′-UTR Constructs into Flp-In Cells

1. Day 1: Seed 3×10^6 Flp-In™ T-REx™ cells into fifty 10 cm tissue culture plates in DMEM supplemented with 10% FBS (*see* **Note 13**).
2. Day 2: 24 h later, transfect cells with 3.75 μg pOG44 and 3 μg pEF5/FRT/V5-D-TOPO® or pcDNA5/FRT containing the dsRed/GFP-3′-UTR expression cassette using Lipofectamine 2000 according to the manufacturer's instructions.
3. Day 3: 24 h post transfection, change the media to DMEM supplemented with 10% FBS and 1% penicillin/streptomycin (*see* **Note 14**).
4. Day 4: Transfer the cells of each 10 cm plate into 4 × 10 cm tissue culture plates using DMEM supplemented with 10% FBS, 1% penicillin/streptomycin, 80–100 μg/ml hygromycin (*see* **Note 2**) and 100 μg/ml zeocin.
5. Start changing the media every 4 days. After a week, colonies of hygromycin-resistant cells will form.
6. After 2–3 weeks, collect all colonies and seed them on a 10 cm tissue culture plate.
7. After 2–3 passages, isolate DNA from 2×10^7 Flp-In™ T-REx™ cells using a DNA extraction kit.
8. Prepare samples for high-throughput sequencing as outlined in **steps 14–16** in Subheading 3.6.

3.8 Fluorescence Activated Cell Sorting (FACS)

1. Seed cells 24 h prior harvesting them for FACS analyses.
2. Wash cells twice with PBS.
3. Trypsinize cells (*see* **Note 15**).
4. Resuspend cells in DMEM growth media.
5. Place a cell strainer on top of a 50 ml Falcon tube and slowly pipette cells through the strainer (*see* **Note 15**).
6. Count cells and confirm that cells are not clumped.
7. Spin cells at $400 \times g$ for 3 min at RT.
8. Wash cells once with PBS.
9. Resuspend cells in cell sorting buffer to a concentration of $\sim 3 \times 10^7$ cells/ml (*see* **Note 3**).
10. Transfer cells to the appropriate FACS tubes.
11. FACS sort of cells based on their GFP expression using, for example, a 488 nm laser and 510/21 bandpass filter for GFP, and a 532 nm laser and 575/25 bandpass filter for dsRed. Determine single cells by their forward and side scatter profiles.
12. Sort dsRed-positive cells with the middle 50% of dsRed intensity (centered on the mode of the distribution).
13. For the identification of regulatory motifs, sort GFP-positive subpopulations (e.g., 0–10%, 20–30%, 40–60%, 70–80%, 90–100% GFP intensity) from the dsRed-positive cells with the middle 50% of dsRed intensity (*see* **Note 16**).
14. Collect at least 0.5–1 million cells.
15. Spin cells and resuspend them in DMEM supplemented with 20% FBS and 1% penicillin/streptomycin (*see* **Note 17**).
16. Plate cells into T25 flasks.
17. 24 h after sorting, change media to DMEM supplemented with 10% FBS, 1% penicillin/streptomycin and 80–100 μg/ml hygromycin (*see* **Note 2**).

3.9 Analyzing Repressive Motifs in Cell SubPopulations

1. Subject cells from each subpopulation with the lower 50% of GFP intensity to DNA isolation and Illumina sequencing (see **steps 14–16** in Subheading 3.6) to analyze enriched motifs in each population.
2. In parallel isolate RNA from 10^6 cells using TRIzol.
3. Perform cDNA synthesis using oligo(dT)-primed RNA.
4. Amplify a region spanning the 3′ end of the GFP-coding sequence and the 3′ end of the 3′-UTR region using Phusion® High-Fidelity DNA Polymerase.
5. Reamplify the surrounding region of the variable nucleotides to add individual barcodes and Illumina sequencing adaptors.
6. Sequence the libraries to obtain 50 nt reads.

7. Calculate enrichment of each motif by dividing the read count (RPM—reads per million) of each motif in the sorted subpopulation normalized to the read count of the whole library by the read count of the motif present in the middle 50% cell population of the dsRed intensity.

3.10 Analyzing Destabilizing Motifs in Cell Subpopulations by Measuring RNA Decay Rates

1. For each time point, seed 5×10^5 Flp-In™ T-REx™ cells into a well of a 6-well plate (*see* **Note 18**).
2. Replace media with DMEM supplemented with 10% FBS, 1% penicillin/streptomycin and 2.5 μg/ml actinomycin D and put cells back into the incubator (*see* **Note 10**).
3. At each time point, remove media and wash cells once with PBS.
4. Lyse cells by adding 1 ml of TRIzol.
5. Proceed to RNA isolation, cDNA synthesis and high throughput Illumina sequencing as mentioned in Subheading 3.9.
6. Calculate the decay rates by counting the reads of each individual motif at the different time points.

4 Notes

1. We recommended fusing the dsRed and GFP coding sequences with endogenous 3′-UTRs (~500–1000 bp) to mimic endogenous gene expression. The sequence motifs of interest are then inserted into the center of the 3′-UTR, which is fused to the GFP open reading frame. Shorter 3′-UTRs that contain only the motifs of interest tend to have higher expression in general and may mask the effect of regulatory motifs. For individual GFP-3′-UTR reporter constructs, a construct with a mutated version of the motif should be generated, preferentially by changing 2–3 nt in the center of the motif.
2. We have observed substantial batch variation in the activity of hygromycin. Before generating stably integrated cell lines with the plasmids pEF5/FRT/V5-D-TOPO® or pcDNA5/FRT, determine the optimal selecting hygromycin concentration for the Flp-In™ T-REx™ cells.
3. This buffer prevents cells from sticking together.
4. Cells should be ~50% confluent.
5. Although media with 10% FBS can be used, a FBS concentration of 30% can result in significantly increased virus yields.
6. We recommend a multiplicity of infection (MOI) of 10 when using shRNA constructs and a MOI of 1–10 for generating cells stably expressing GFP-3′-UTR reporter constructs.

7. Leave plates on side and remove all excess PBS.
8. At this step the cell lysate can be snap frozen and stored at −80 °C, but is it recommended to proceed with the fresh lysate.
9. mRNAs containing motifs that promote translation will undergo translation by relatively increased numbers of ribosomes and will therefore sediment into fractions with higher sucrose concentration.
10. DMEM supplemented with actinomycin should be prewarmed to 37 °C prior to use.
11. In mammals, the majority of mRNAs are initially degraded via poly(A) tail shortening mediated by deadenylases, such as the CCR4-NOT and PAN2-PAN3 complex or PARN. Depletion of these deadenylases will result in increased levels of GFP-3′-UTR reporter mRNAs containing the wild-type motif in comparison to a mutated version, when analyzed in control-depleted cells and cells depleted for specific decay factors. Following deadenylation, mRNAs are further degraded via 5′–3′ or 3′–5′ mRNA decay pathways. Increased mRNA levels in cells depleted for the decapping complex and/or XRN1 will reveal whether mRNAs are degraded via the 5′–3′ decay pathway. In contrast, increased reporter mRNA levels due to the depletion of exosome components or proteins from the DIS3L family will indicate the degradation via the 3′–5′ mRNA decay pathway. When decay enzymes of both pathways (5′–3′ or 3′–5′ direction) are depleted, deadenylated transcripts accumulate in the cell. Therefore cDNA synthesis is performed with gene-specific primers, rather than dT_n primers.
12. During oligonucleotide synthesis, each nucleotide has a different coupling efficiency and therefore the incorporation rate varies between the A, T, C, and G nucleotides. To ensure that there is an equal distribution of nucleotides for the degenerated positions, specify in the order form for the oligonucleotide that the nucleotides for the degenerated position are applied as "hand-mixed." For example, if degenerate positions should have an equal distribution of all four nucleotides, indicate that the occurrence of each nucleotide has to be 25% at each position.
13. Keep some of the cells and split them on a regular basis. They will be used as a baseline for FACS measurements later. Also, transfect cells with constructs expressing either the GFP or dsRed fluorophores. These cells will be used for fluorescence compensation due to spectral overlap of the fluorophores, although there is little overlap between the suggested settings when GFP and dsRED as fluorophores are used. The number of 10 cm tissue culture plates may vary depending on the

efficiency of generating stably integrated cell clones. It is recommended to test first in a pilot experiment, starting with one 10 cm tissue culture plate, how many plates are needed to obtain a representative coverage of all randomized sequences in the library.

14. At this point, addition of the selecting antibiotic zeocin to the media is appropriate, but not essential.
15. It is important that the cells for FACS analyses and sorting are in single-cell suspension.
16. It is essential to sort and culture cells from the middle 50% of GFP intensity to identify enriched motifs in the subpopulations.
17. It is gentler to the cells not to add hygromycin at this point.
18. Seed also cells from the middle 50% GFP intensity of the FACS sort to analyze the decay rates of GFP-3′-UTR constructs with "neutral motifs." Furthermore, for the identification of mRNA destabilizing motifs it is essential to analyze the RNA decay rates, since motifs that reduce e.g., translation rates will also have reduced GFP intensities.

Acknowledgment

This work was supported by a Research Scholar Grant from the American Cancer Society and R01GM105668 from NIGMS (both to A.G.).

References

1. Roy B, Jacobson A (2013) The intimate relationships of mRNA decay and translation. Trends Genet 29:691–699
2. Garneau NL, Wilusz J, Wilusz CJ (2007) The highways and byways of mRNA decay. Nat Rev Mol Cell Biol 8:113–126
3. Siepel A, Bejerano G, Pedersen JS, Hinrichs AS, Hou M, Rosenbloom K, Clawson H, Spieth J, Hillier LW, Richards S, Weinstock GM, Wilson RK, Gibbs RA, Kent WJ, Miller W, Haussler D (2005) Evolutionarily conserved elements in vertebrate, insect, worm, and yeast genomes. Genome Res 15:1034–1050
4. Xie X, Lu J, Kulbokas EJ, Golub TR, Mootha V, Lindblad-Toh K, Lander ES, Kellis M (2005) Systematic discovery of regulatory motifs in human promoters and 3′ UTRs by comparison of several mammals. Nature 434:338–345
5. Andreassi C, Riccio A (2009) To localize or not to localize: mRNA fate is in 3′UTR ends. Trends Cell Biol 19:465–474
6. Balagopal V, Parker R (2009) Polysomes, P bodies and stress granules: states and fates of eukaryotic mRNAs. Curr Opin Cell Biol 21:403–408
7. Dreyfuss G, Kim VN, Kataoka N (2002) Messenger-RNA-binding proteins and the messages they carry. Nat Rev Mol Cell Biol 3:195–205
8. Kuersten S, Goodwin EB (2003) The power of the 3′ UTR: translational control and development. Nat Rev Genet 4:626–637
9. Mazumder B, Seshadri V, Fox PL (2003) Translational control by the 3′-UTR: the ends specify the means. Trends Biochem Sci 28:91–98
10. Mitchell P, Tollervey D (2000) mRNA stability in eukaryotes. Curr Opin Genet Dev 10:193–198

11. Moore MJ (2005) From birth to death: the complex lives of eukaryotic mRNAs. Science 309:1514–1518
12. Palacios IM, St Johnston D (2001) Getting the message across: the intracellular localization of mRNAs in higher eukaryotes. Annu Rev Cell Dev Biol 17:569–614
13. Bhandari D, Raisch T, Weichenrieder O, Jonas S, Izaurralde E (2014) Structural basis for the Nanos-mediated recruitment of the CCR4-NOT complex and translational repression. Genes Dev 28:888–901
14. Braun JE, Huntzinger E, Fauser M, Izaurralde E (2011) GW182 proteins directly recruit cytoplasmic deadenylase complexes to miRNA targets. Mol Cell 44:120–133
15. Chekulaeva M, Mathys H, Zipprich JT, Attig J, Colic M, Parker R, Filipowicz W (2011) miRNA repression involves GW182-mediated recruitment of CCR4-NOT through conserved W-containing motifs. Nat Struct Mol Biol 18:1218–1226
16. Collart MA (2016) The Ccr4-Not complex is a key regulator of eukaryotic gene expression. Wiley Interdiscip Rev RNA 7:438–454. https://doi.org/10.1002/wrna.1332
17. Lykke-Andersen J, Wagner E (2005) Recruitment and activation of mRNA decay enzymes by two ARE-mediated decay activation domains in the proteins TTP and BRF-1. Genes Dev 19:351–361
18. Leppek K, Schott J, Reitter S, Poetz F, Hammond MC, Stoecklin G (2013) Roquin promotes constitutive mRNA decay via a conserved class of stem-loop recognition motifs. Cell 153:869–881
19. Van Etten J, Schagat TL, Hrit J, Weidmann CA, Brumbaugh J, Coon JJ, Goldstrohm AC (2012) Human Pumilio proteins recruit multiple deadenylases to efficiently repress messenger RNAs. J Biol Chem 287:36370–36383
20. Geissler R, Simkin A, Floss D, Patel R, Fogarty EA, Scheller J, Grimson A (2016) A widespread sequence-specific mRNA decay pathway mediated by hnRNPs A1 and A2/B1. Genes Dev 30:1070–1085
21. Chen CY, Gherzi R, Ong SE, Chan EL, Raijmakers R, Pruijn GJ, Stoecklin G, Moroni C, Mann M, Karin M (2001) AU binding proteins recruit the exosome to degrade ARE-containing mRNAs. Cell 107:451–464
22. Bartel DP (2009) MicroRNAs: target recognition and regulatory functions. Cell 136:215–233
23. Barreau C, Paillard L, Osborne HB (2005) AU-rich elements and associated factors: are there unifying principles? Nucleic Acids Res 33:7138–7150
24. Staals RH, Bronkhorst AW, Schilders G, Slomovic S, Schuster G, Heck AJ, Raijmakers R, Pruijn GJ (2010) Dis3-like 1: a novel exoribonuclease associated with the human exosome. EMBO J 29:2358–2367
25. Tomecki R, Kristiansen MS, Lykke-Andersen S, Chlebowski A, Larsen KM, Szczesny RJ, Drazkowska K, Pastula A, Andersen JS, Stepien PP, Dziembowski A, Jensen TH (2010) The human core exosome interacts with differentially localized processive RNases: hDIS3 and hDIS3L. EMBO J 29:2342–2357
26. Parker R, Sheth U (2007) P bodies and the control of mRNA translation and degradation. Mol Cell 25:635–646
27. Malecki M, Viegas SC, Carneiro T, Golik P, Dressaire C, Ferreira MG, Arraiano CM (2013) The exoribonuclease Dis3L2 defines a novel eukaryotic RNA degradation pathway. EMBO J 32:1842–1854
28. Lubas M, Damgaard CK, Tomecki R, Cysewski D, Jensen TH, Dziembowski A (2013) Exonuclease hDIS3L2 specifies an exosome-independent 3′-5′ degradation pathway of human cytoplasmic mRNA. EMBO J 32:1855–1868
29. Chaudhury A, Kongchan N, Gengler JP, Mohanty V, Christiansen AE, Fachini JM, Martin JF, Neilson JR (2014) A piggyBac-based reporter system for scalable in vitro and in vivo analysis of 3′ untranslated region-mediated gene regulation. Nucleic Acids Res 42:e86
30. Weingarten-Gabbay S, Elias-Kirma S, Nir R, Gritsenko A, Stern-Ginossar N, Yakhini Z, Weinberger A, Segal E (2016) Comparative genetics. Systematic discovery of cap-independent translation sequences in human and viral genomes. Science 351: aad4939
31. Wissink EM, Fogarty EA, Grimson A (2016) High-throughput discovery of post-transcriptional cis-regulatory elements. BMC Genomics 17:177
32. Chen CY, Ezzeddine N, Shyu AB (2008) Messenger RNA half-life in mammalian cells. Methods Enzymol 448:335–357

Chapter 7

iCLIP of the PIWI Protein Aubergine in *Drosophila* Embryos

Bridlin Barckmann, Jérémy Dufourt, and Martine Simonelig

Abstract

Piwi-interacting RNAs (piRNAs) are a class of small noncoding RNAs bound to specific Argonaute proteins, the PIWI proteins. piRNAs target mRNAs by complementarity to silence them; they play an important role in the repression of transposable elements in the germ line of many species. piRNAs and PIWI proteins are also involved in diverse biological processes through their role in the regulation of cellular mRNAs. In the *Drosophila* embryo, they contribute to the maternal mRNA decay occurring during the maternal-to-zygotic transition. CLIP (UV cross-linking and immunoprecipitation) techniques have been used to identify target mRNAs of Argonaute proteins. Here we describe the iCLIP (individual-nucleotide resolution CLIP) protocol that we have adapted for the PIWI protein Aubergine in *Drosophila* embryos.

Key words Aubergine, CLIPseq, *Drosophila*, Embryo, iCLIP, Immunoprecipitation, Maternal-to-zygotic transition, mRNA decay, NGS, piRNA, PIWI protein, Reverse transcription, UV cross-link

1 Introduction

piRNAs are a specific class of small, 23–30 nucleotides (nt), non-coding RNAs interacting with the Argonaute proteins of the PIWI clade [1]. An important function of the piRNA pathway is the repression of transposable elements in the germ line. piRNAs target transposable element mRNAs through complementarity and guide their cleavage by the cytoplasmic PIWI proteins Aubergine (Aub) and Argonaute 3 (Ago3) bound to these piRNAs. A major question in the field has been whether piRNAs are also involved in silencing cellular (non-transposable element) mRNAs.

The maternal-to-zygotic transition is a conserved developmental process during which the control of embryonic development switches from the maternal to the zygotic genome. It involves the massive degradation of maternal mRNAs, in addition to activation of the zygotic genome [2]. We have shown that the piRNA pathway is involved in the decay of the maternal *nanos* mRNA during the maternal-to-zygotic transition [3]. *nanos* encodes a conserved protein required for germ cell development. piRNAs produced from

Shireen R. Lamandé (ed.), *mRNA Decay: Methods and Protocols*, Methods in Molecular Biology, vol. 1720, https://doi.org/10.1007/978-1-4939-7540-2_7,

transposable element sequences target *nanos* mRNA 3′UTR and guide interactions with Aub and Ago3, which in turn contribute to *nanos* mRNA deadenylation and decay by recruiting the CCR4-NOT deadenylation complex [3]. This regulation is essential for embryonic development. Thus, this study identified the first example of piRNA-dependent mRNA regulation with a key biological function. Later studies revealed that piRNA-dependent mRNA silencing is a widespread and conserved mechanism of gene regulation. piRNAs are involved in massive degradation of mRNAs during mouse sperm development and maturation [4–7]. This mRNA decay depends on the mouse PIWI protein, Miwi, and occurs either through the recruitment of CAF1, a deadenylase of the CCR4-NOT deadenylation complex, or through direct cleavage by Miwi. Another example showed that piRNA-dependent decay of an mRNA encoding a masculinization factor is the primary determinant of sex determination in *Bombyx mori* [8]. These results establish the role of piRNAs in posttranscriptional regulation of gene expression.

High-throughput sequencing following UV cross-linking and immunoprecipitation (HITS-CLIP or CLIPseq) methods have been developed to identify binding sites of RNA binding proteins on a genome-wide scale [9]. HITS-CLIP identifies protein–RNA interaction sites by cross-linking proteins to bound RNA molecules, purifying these protein–RNA complexes, and sequencing the associated RNAs. This powerful technique was then applied to Argonaute proteins to identify mRNAs interacting with microRNAs (miRNAs)-loaded Argonaute and their binding sites [10–12]. The technique has the potential to reveal both Argonaute–miRNA interactions and Argonaute-mRNA binding sites. It provided significant progress towards establishing high resolution maps of miRNA-mediated mRNA regulation at genome-wide level, which can be used for functional validation. CLIP-based approaches were not designed to identify the specific miRNAs that are responsible for specific Argonaute–mRNA interactions. However, further bioinformatic analyses of data sets, as well as developments of HITS-CLIP techniques allow identification of chimeric mRNA–miRNA molecules, which represent Argonaute–-miRNA–mRNA ternary complexes and unambiguously identify the miRNA linked to a specific target site (*see* [13] for Review).

Another technical development was set up with iCLIP (individual-nucleotide resolution CLIP), which improves RNA–protein binding resolution to the nt level [14]. iCLIP is based on the fact that the cross-link sites largely block reverse transcriptase progression [15]. Thus, the truncated cDNAs are recovered and sequenced as a mean to identify the exact sites, at the nt level, of RNA–protein interactions (Fig. 1). iCLIP also uses barcoded primers for cDNA library production, to remove PCR duplicates arising from PCR amplification of libraries. It has been used to address Argonaute/miRNA-dependent mRNA regulation [16–18].

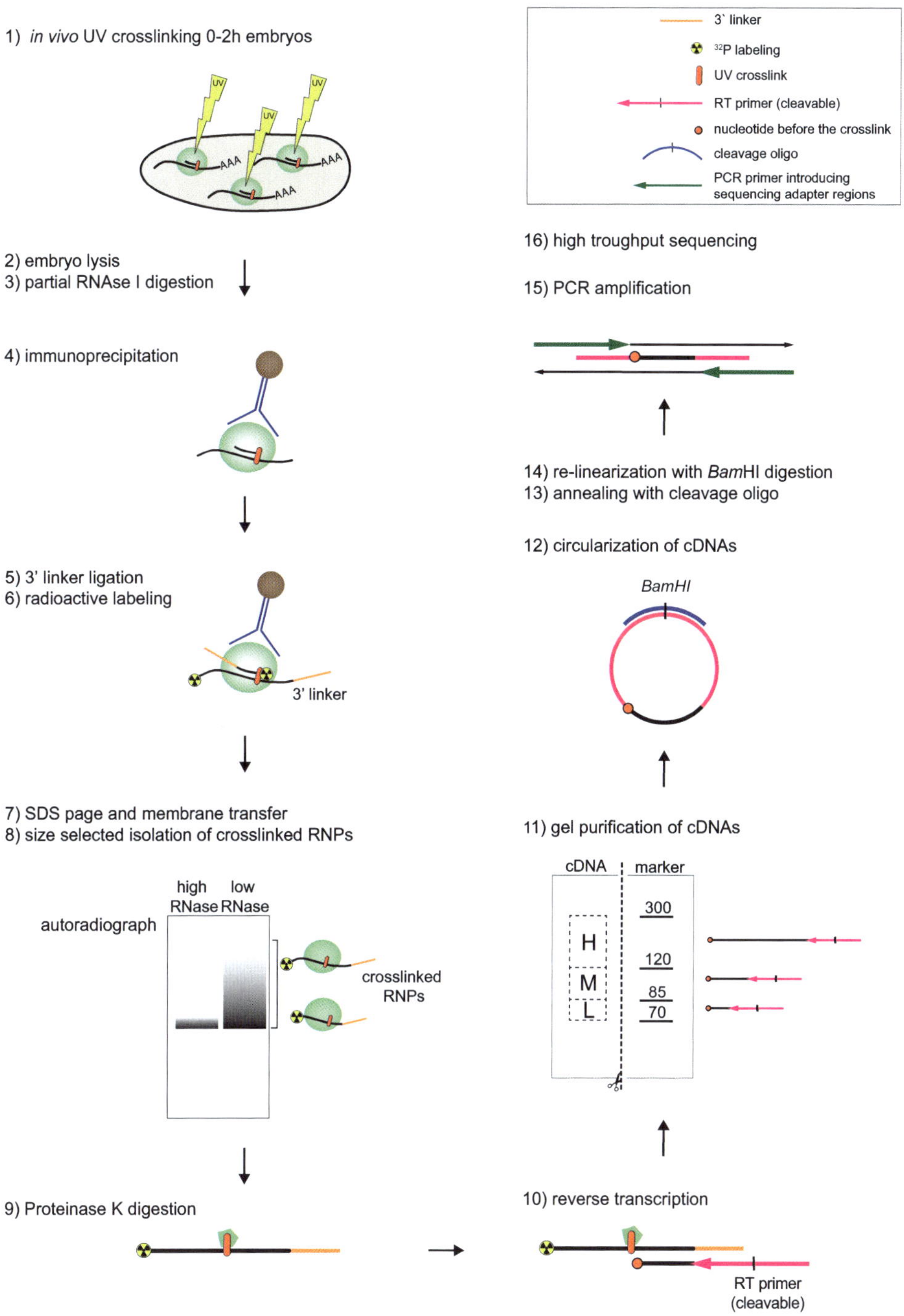

Fig. 1 Schematic overview of the iCLIP experimental protocol. 0–2-h-old embryos are UV cross-linked, which induces covalent binding (*orange ovals*) between interacting proteins and RNAs. The embryos are lysed and the RNA partially digested. RNPs with the protein of interest are isolated using antibody conjugated to magnetic beads. The RNA in the RNPs is ligated to a linker at the 3′-end (3′ linker) and radioactively labeled with ^{32}P at the 5′-end. After elution from the beads the RNPs are separated on a SDS-PAGE gel and transferred to nitrocellulose membrane. The RNPs from the low RNase I condition are isolated from the membrane and

CLIP-based techniques also played a critical role in identifying mRNA targets of PIWI proteins and establishing the function of PIWI proteins in cellular mRNA regulation. HITS-CLIP of the PIWI proteins, Miwi and Mili, were performed in mouse testes and round spermatids [7, 19]. Piwi HITS-CLIP was also performed in *Drosophila* ovarian somatic cells in culture [20].

We performed iCLIP of Aub in *Drosophila* embryos to address the generality of Aub-dependent maternal mRNA regulation [21]. This approach identified several hundred maternal mRNAs bound by Aub. For approximately two hundreds of these mRNAs, Aub interaction led to their decay during the maternal-to-zygotic transition. Interestingly, Aub-dependent unstable mRNAs were enriched in mRNAs encoding germ cell determinants; these mRNAs were indeed degraded in the bulk of the embryo, which corresponds to the somatic region, but they were stabilized in the future germ line. iCLIP with unloaded Aub revealed that Aub interaction with mRNAs required its loading with piRNAs. Aub-dependent mRNA decay occurred partly through cleavage via the endonuclease activity of Aub [21]. Aub binding to mRNAs encoding germ cell determinants was also found in an independent study using Aub HITS-CLIP in embryos [22].

Here, we describe the adaptation of the iCLIP protocol from Konig et al. [23] for use in *Drosophila* embryos (Fig. 1). The technique was set up with a GFP-tagged version of Aub, therefore this protocol can be applied to any GFP-tagged RNA binding protein.

2 Materials

2.1 Embryo Collection

1. Embryo collection cages (fitting to 60 mm petri dishes).
2. Red fly food plates. Sixty millimeter petri dishes filled with regular fly food stained with neutral red, to help visualizing embryos.

Fig. 1 (continued) treated with Proteinase K to digest the proteins from the RNA. The protein digestion leaves a residual peptide at the cross-link site, which sterically hinders the reverse transcriptase during cDNA production. Most cDNAs stop at the nt (*red circle*) before the cross-link site. The RT (reverse transcription) primer that anneals to the 3′ linker introduces a cleavable site, as well as an experiment specific barcode, and a random barcode, for bioinformatic analyses of sequences. cDNAs are size selected on a gel; the marker that is run on the same gel is used to determine the areas containing the cDNAs of interest, which must be cut out. The low cut (L) is between 70 and 85 nt, the medium cut (M) between 85 and 120 nt, and the high cut (H) between 120 and 300 nt. After eluting from the gel, the cDNAs are circularized and an oligonucleotide is annealed to the cleavable site to allow *Bam*HI digestion. cDNA circles are relinearized to form cDNA molecules with a linker at both ends, which can be used for PCR amplification

3. Yeast paste: dry baker's yeast mixed with water until it forms a paste that should be not too watery to prevent the flies from sticking to it.
4. Fifty percent bleach solution using sodium hypochlorite (6–14% active chlorine).
5. Sieve for embryo washings: Made out of a 50 ml falcon tube, with a cut open lid and cut open bottom, and a 70 μm pore sized mesh that is screwed with the opened lid to the falcon tube to form a sieve.
6. Soft paintbrush.
7. Wash bottle.
8. Distilled water.

2.2 UV Cross-Linking

1. Freshly collected 0–2 h embryos.
2. 10 ml reaction tube.
3. Ice-cold 0.1% Tween.
4. Ice bucket.
5. 100 mm petri dishes.
6. Soft paint brush.
7. UV cross-linker (BIO-LINK type BLX with five 8 W tubes).
8. Ice-cold 1× PBS (Sigma # D8537-500 ml).
9. 1.5 ml reaction tubes.
10. Wash bottle.
11. Dry ice.

2.3 Embryo Lysis, Partial RNA Digestion, Immunoprecipitation

2.3.1 Bead Preparation

1. Protein A Dynabeads (Thermo Fisher/Invitrogen).
2. Lysis buffer (without protease inhibitor and ANTI-RNase): 50 mM Tris–HCl (pH 7.4), 100 mM NaCl, 1% NP-40, 0.1% SDS, 0.5% sodium deoxycholate.
3. α-GFP polyclonal rabbit serum (Thermo Fisher/Invitrogen).
4. Nonimmune rabbit serum.
5. Rotation wheel at room temperature.

2.3.2 Embryo Lysis, Partial RNA Digestion, and Immunoprecipitation

1. Protease inhibitor Cocktail III (Calbiochem).
2. ANTI-RNase (Ambion).
3. Lysis buffer plus protease inhibitor and ANTI-RNase: 10 μl Protease inhibitor Cocktail III and 1 μl ANTI-RNase per 1 ml lysis buffer (from Subheading 2.3.1).
4. RNase I (Ambion).
5. Turbo DNase (Ambion).
6. Thermomixer (Eppendorf).

7. Centrifuge.
8. 1.5 ml reaction tubes.
9. High salt buffer: 50 mM Tris–HCl (pH 7.4), 1 M NaCl, 1 mM EDTA, 1% NP-40, 0.1% SDS, 0.5% sodium deoxycholate.
10. PNK wash buffer: 20 mM Tris–HCl (pH 7.4), 10 mM $MgCl_2$, 0.2% Tween 20.
11. Rotation wheel at 4 °C.
12. Bucket with ice.

2.4 Linker Ligation to RNA 3′-Ends

1. H_2O (*see* **Note 1**).
2. 10× ligation Buffer: 500 mM Tris pH 7.5, 100 mM $MgCl_2$, 600 μg/ml BSA, 100 mM DTT (can be prepared in advance and kept at −20 °C in aliquots).
3. DMSO (Sigma).
4. RNAsin (Promega).
5. Truncated T4 RNA ligase 2 (Rnl2) (New England Biolabs).
6. Preadenylated linker (IDT, miRNA cloning linker2): 5′App-CACTCGGGCACCAAGGA-ddC.

   (5′App: 5′adenylation; ddC: dideoxy-C to block the 3′-end) (Fig. 2).

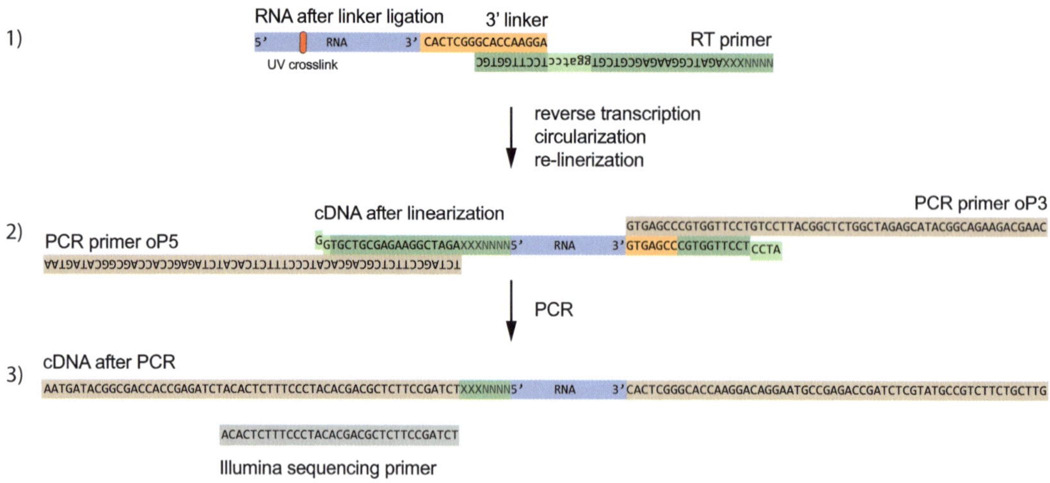

Fig. 2 Schematic representation of the primers used in the iCLIP. (1) The RNA molecule (*blue*) is ligated to a DNA linker (3′ linker, *orange*) at its 3′-end. Then, the RT primer (*green*) is annealed to the first 10 nt of the 3′ linker for reverse transcription. The RT primer introduces a *Bam*HI restriction enzyme site (*light green*, *small letters*), a random barcode (NNNN), and an experiment specific barcode (XXX). (2) The cDNA is shown after reverse transcription, circularization and relinearization. The PCR primers oP3 and oP5 are shown in beige; they introduce sequences that are required for the sequencing. (3) The cDNA molecule after PCR is shown along with the paired end sequencing primer (*grey*) that is required for sequencing

7. PNK wash buffer (from Subheading 2.3.2).
8. High salt buffer (from Subheading 2.3.2).
9. Rotation wheel at 4 °C.

2.5 Radioactive Labeling

1. H_2O.
2. PNK buffer (provided with the T4 Polynucleotide Kinase 3′ phosphatase minus).
3. ^{32}P-γ-ATP (3000 Ci/mmol, 10 mCi/ml, Perkin Elmer).
4. T4 Polynucleotide Kinase 3′ phosphatase minus (New England Biolabs).
5. Thermomixer (Eppendorf).
6. Nupage loading Buffer (Thermo Fisher/Invitrogen).

2.6 SDS-PAGE and Membrane Transfer

1. NuPAGE® Novex 4–12% Bis-Tris Gel (Thermo Fisher/Invitrogen).
2. Nitrocellulose membrane (Whatman).
3. Wet blot system, that fits the NuPAGE® Novex 4–12% Bis-Tris Gels.
4. NuPAGE® Transfer Buffer (Thermo Fisher/Invitrogen).
5. Film (Kodak Biomax MR).
6. Film cassette.

2.7 Ribonucleoprotein (RNP) Isolation and Proteinase K Digestion

1. Needle.
2. Scalpel blade.
3. 1.5 ml reaction tubes.
4. Thermomixer (Eppendorf).
5. PK buffer: 100 mM Tris–HCl (pH 7.4), 50 mM NaCl, 10 mM EDTA.
6. Proteinase K (Roche).
7. PK urea buffer: 100 mM Tris–HCl (pH 7.4), 50 mM NaCl, 10 mM EDTA, 7 M urea.
8. RNA phenol/$CHCl_3$ (Ambion).
9. 2 ml Phase Lock Gel heavy tubes (VWR).
10. Centrifuge.
11. Glycoblue (Ambion).
12. 3 M sodium acetate (pH 5.5) (Ambion).
13. Ice-cold 100% ethanol.

2.8 cDNA Production

2.8.1 Reverse Transcription (RT)

1. H_2O.
2. Centrifuge.
3. 80% ethanol.
4. RT primer (0.5 pmol/μl) (with a unique experiment specific barcode for each reaction):

   5′P-NNNNXXXAGATCGGAAGAGCGTCGTggatccTCCTTGGTGC
   (N: random barcode, where N is any nt; X: experiment specific barcode; 5′P: 5′ phosphate; ggatcc: *Bam*HI restriction site) (Fig. 2).
5. dNTPs (10 mM).
6. 5× RT buffer (provided with Superscript III reverse transcriptase).
7. 0.1 M DTT.
8. Superscript III reverse transcriptase (Thermo Fisher/Invitrogen).
9. PCR cycler.
10. Glycoblue (Ambion).
11. 3 M sodium acetate (pH 5.5) (Ambion).
12. Ice-cold 100% ethanol.

2.8.2 Gel Purification of cDNAs

1. H_2O.
2. Centrifuge.
3. 80% ethanol.
4. 2× TBE-Urea loading buffer (Bio-Rad).
5. Heat block.
6. 5% TBE-Urea Gel (Bio-Rad).
7. Low molecular weight ladder (New England Biolabs).
8. SYBR Gold Nucleic Acid Gel Stain (Thermo Fisher/Invitrogen).
9. 1.5 ml reaction tubes.
10. TE buffer.
11. Plastic pestles, fitting 1.5 ml reaction tubes.
12. Thermomixer (Eppendorf).
13. Costar Spin X column (Coring incorporated).
14. 1 cm glass pre filters (Whatman).
15. Glycoblue (Ambion).
16. 3 M sodium acetate (pH 5.5) (Ambion).
17. Ice-cold 100% ethanol.

2.9 Circularization and Relinearization of cDNAs

1. H_2O.
2. Centrifuge.
3. 80% ethanol.
4. 10× CircLigase II buffer (provided with CircLigase II).
5. 50 mM $MnCl_2$.
6. CircLigase II (Epicentre).
7. Heat block.
8. 10× FastDigest buffer (provided with Fermentas FastDigest *Bam*HI).
9. 10 μM oligoAN: CAAGGAGGATCCACGACGCTCTTCaaaa.
10. FastDigest *Bam*HI (Fermentas).
11. Glycoblue (Ambion).
12. 3 M sodium acetate (pH 5.5) (Ambion).
13. Ice-cold 100% ethanol.

2.10 PCR and Preparation for Deep Sequencing

1. H_2O.
2. Centrifuge.
3. 80% ethanol.
4. oP3/oP5 10 μM oligo mix:

   oP3: CAAGCAGAAGACGGCATACGAGATCGGTCTCGG CATTCCTGTCCTTGGTGCCCGAGTG (Fig. 2).
   oP5: AATGATACGGCGACCACCGAGATCTACACTCTTT CCCTACACGACGCTCTTCCGATCT (Fig. 2).
5. Accuprime Supermix 1 (Thermo Fisher/Invitrogen).
6. PCR cycler.
7. 2× TBE-Urea loading buffer (Bio-Rad).
8. 5% TBE-Urea Gel (Bio-Rad).
9. Low molecular weight ladder (New England Biolabs).
10. SYBR Gold Nucleic Acid Gel Stain (Thermo Fisher/Invitrogen).
11. Agilent analyzer.

3 Methods

3.1 Embryo Collection

1. Embryo collections are obtained from laying cages closed with a 60 mm petri dish filled with red fly food with a dab of baker's yeast (*see* **Note 2**). To obtain 0–2 h embryos let the flies lay for 2 h at 25 °C. The embryos are then washed from the laying plate into a sieve placed into a 200 ml beaker, using a wash

bottle with water and a soft paintbrush. Wash the embryos with water until they are free of yeast and fly food.

2. Embryos are dechorionated by placing the sieve containing the embryos for 2 min into a 200 ml glass beaker containing 50% bleach (*see* **Note 3**).
3. Rinse the embryos thoroughly with distilled water.

3.2 UV Cross-Linking [24]

1. Transfer the dechorionated embryos from the sieve into a 10 ml reaction tube filled with ice-cold 0.1% Tween 20 and vortex them briefly (5–10 s) to separate the embryos from each other then transfer them into a petri dish on ice (*see* **Note 4**).
2. UV cross-link on ice with six times 5 min pulses at 254 nm, in a UV cross linker. Stir the embryos between each UV pulse and go on to the next UV pulse without recovery pause (*see* **Note 5**).
3. Wash the embryos from the petri dish with a washing bottle filled with ice-cold PBS into a sieve.
4. Wash with ice-cold PBS until all the Tween 20 is removed and transfer them to a 1.5 ml reaction tube filled with ice-cold PBS and remove all PBS (*see* **Note 6**).
5. Freeze on dry ice and store at −80 °C.

3.3 Embryo Lysis, Partial RNA Digestion, Immunoprecipitation

3.3.1 Bead Preparation

Use 1 ml of buffer for each bead washing step and 100 μl of Protein A Dynabead slurry for each immunoprecipitation.

1. Transfer the Dynabead slurry into a fresh 1.5 ml reaction tube.
2. Wash three times 5 min with lysis buffer at room temperature (*see* **Note 7**).
3. Resuspend the beads in 100 μl lysis buffer and add 1–5 μg antibody: 6 μl α-GFP polyclonal rabbit serum for 100 μl embryos, or 6 μl rabbit serum for the no-antibody control (*see* **Note 8**).
4. Rotate the tubes at room temperature for 1 h.
5. Wash three times 5 min with lysis buffer. Leave in lysis buffer until the embryo lysates are ready.

3.3.2 Embryo Lysis, Partial RNA Digestion, and Immunoprecipitation

All steps are performed on ice or at 4 °C unless indicated otherwise. To protect the RNA, all steps must be performed in RNase-free conditions.

1. Crush 100 μl embryos in 200 μl lysis buffer plus protease inhibitor and ANTI-RNase with a plastic pestle. After crushing the embryos make up to 1 ml with lysis buffer plus protease inhibitor and ANTI-RNase.

2. Prepare a 1:3000 RNase I dilution for the actual experiment in low RNase conditions and a 1:50 RNase I dilution for the high RNase control experiment (*see* **Note 9**).
3. Add 10 μl of the appropriate RNase I dilution and 5 μl Turbo DNase to 1 ml lysate.
4. Incubate for 3 min at 37 °C in in the thermomixer at 1100 rpm. Transfer immediately to ice.
5. Spin the lysates three times at 16,000 × *g* at 4 °C for 10 min to clear the lysates. Each time transfer the cleared lysate into a fresh 1.5 ml reaction tube without transferring the pellet or the white layer on top of the supernatant.
6. Save 15 μl as input control.
7. Remove the lysis buffer from the prepared beads and add the lysate.
8. Rotate the samples at 4 °C overnight.
9. Save 15 μl of the supernatant as depletion control.
10. Discard the supernatant.
11. Wash the beads two times for 5 min with 1 ml high salt buffer at 4 °C on a wheel.
12. Wash the beads two times for 5 min with 1 ml PNK wash buffer at 4 °C on a wheel.

3.4 Linker Ligation to RNA 3′-Ends (on Beads)

1. Remove the supernatant from the beads and resuspend the beads in 20 μl ligation mix:
   - 12 μl H_2O.
   - 2 μl 10× ligation buffer.
   - 2 μl DMSO.
   - 1 μl RNAsin.
   - 2 μl truncated T4 Rnl2.
   - 1 μl preadenylated linker.
2. Incubate for 3 h at 25 °C in the thermomixer at 1000 rpm (*see* **Note 10**).
3. Add 500 μl PNK wash buffer.
4. Wash the beads two times 5 min with 1 ml high salt buffer at 4 °C on a wheel.
5. Wash the beads two times 5 min with 1 ml PNK wash buffer at 4 °C on a wheel (*see* **Note 11**).

3.5 Radioactive Labeling (on Beads)

1. Remove the supernatant and add 8 μl PNK labeling mix containing:
   - 6 μl H_2O.
   - 0.8 μl PNK buffer.
   - 0.8 μl ^{32}P-γ-ATP.
   - 0.4 μl T4 Polynucleotide Kinase 3′ phosphatase minus.
2. Incubate for 5 min at 37 °C in the thermomixer at 1000 rpm.
3. Remove the radioactive PNK labeling mix and add 20 μl Nupage loading Buffer.
4. Incubate for 10 min at 70 °C in the thermomixer at 1000 rpm to elute the RNPs.

3.6 SDS-PAGE and Membrane Transfer

1. Load the eluate on a NuPAGE® Novex 4–12% Bis-Tris gel and run the gel at 4 °C for 2 h at 180 V (depending on the size of the protein).
2. Transfer the RNP complexes onto a nitrocellulose membrane in a wet blot system using NuPAGE® Transfer Buffer for either 3 h at 180 V or overnight at 50 V, under cooled conditions (*see* **Note 12**).
3. Rinse the membrane in PBS and wrap it in plastic foil (heat sealable bag) with a fluorescent sticker (*see* **Note 13**).
4. Expose the membrane to a film for either 30 min, or 1 h, or overnight (Fig. 3a) (*see* **Note 14**).

3.7 RNP Isolation and Proteinase K Digestion

1. Cut out the membrane piece that contains the RNP complexes from the low RNase experiment using the autoradiograph as a mask (*see* **Note 15**). Use a needle to mark the corners of the region of interest on the membrane through the autoradiograph. Then, use a clean scalpel to cut out the membrane piece.
2. Cut the membrane piece into smaller pieces and place them in a 1.5 ml reaction tube with 200 μl PK buffer and 10 μl Proteinase K. Incubate for 20 min at 37 °C in the thermomixer at 1000 rpm.
3. Add 200 μl PK urea buffer.
4. Incubate for 20 min at 37 °C in the thermomixer at 1000 rpm.
5. Transfer the solution without the membrane pieces into a 2 ml Phase Lock Gel heavy tube, add 400 μl RNA phenol/$CHCl_3$. Incubate for 5 min at 30 °C in the thermomixer at 1000 rpm. Separate the phases by spinning 10 min at 16,000 × *g* in a centrifuge at room temperature (*see* **Note 16**).
6. Transfer the aqueous phase into a new tube. Add 0.5 μl glycoblue and 40 μl 3 M sodium acetate (pH 5.5) mix the solution and add 1 ml ice-cold ethanol. Mix again and precipitate for 1 h to overnight at −20 °C.

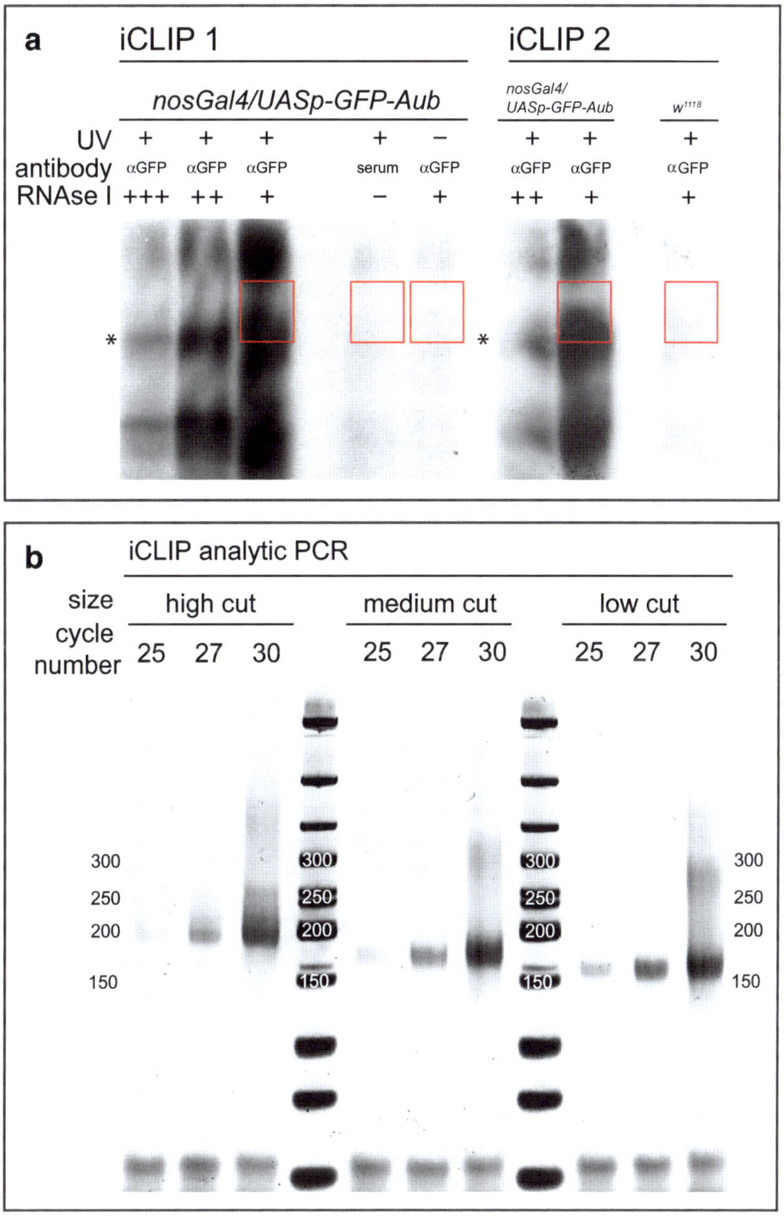

Fig. 3 Examples of autoradiographs and analytical PCR in iCLIP. (**a**) Autoradiographs of ^{32}P-labeled RNPs for two GFP-Aub iCLIP experiments. In high RNase conditions (+++ and ++) the RNP complexes have a discrete size, in the form of a sharp (+++) or a fuzzy (++) band, around the size of the protein alone (*asterisk*). In the low RNase condition (+) the RNPs form a broader band starting from the size of the band in the high RNase condition (+++). The negative controls, no antibody (iCLIP 1, lane 4), no UV (iCLIP 1, lane 5), and no antigen (iCLIP 2, lane 3) are shown. The iCLIP should result in a single band on the autoradiograph, corresponding to the immunoprecipitated protein. In Aub iCLIPs, we always had two additional bands migrating on both sides of Aub size. Because both bands were revealed with Aub antibody on western blots, we believe that they correspond to different forms of Aub (possibly dimers and degradation products). (**b**) Example of an analytic PCR during GFP-Aub iCLIP. The PCR reaction is performed separately for the high, medium and low cut cDNA fractions. The PCR reactions are analyzed after 25, 27 and 30 cycles to determine the optimal cycle number

3.8 cDNA Production

3.8.1 Reverse Transcription (See Note 17)

1. Spin for 30 min at 4 °C at 16,000 × *g*. Remove the supernatant and wash with 80% ethanol (*see* **Note 18**).
2. Spin for 5 min at 4 °C at 16,000 × *g*. Remove ethanol and spin briefly again to completely remove ethanol (*see* **Note 19**).
3. Resuspend the pellet in 6.25 μl H_2O.
4. Transfer the RNA solution in a tube fitting the thermocycler.
5. Add 0.5 μl of RT primer (0.5 pmol/μl) and 0.5 μl of dNTPs (10 mM). For each experiment use a RT primer with a unique experiment-specific barcode (*see* **Note 20**).
6. Incubate for 5 min at 70 °C and cool down to 28 °C in the thermocycler.
7. To each reaction, add in the thermocycler, 2.75 μl of RT mix consisting of:
   - 2 μl 5× RT buffer.
   - 0.5 μl 0.1 M DTT.
   - 0.25 μl Superscript III reverse transcriptase.
8. Mix briefly and spin down to collect the liquid at the bottom of the tube.
9. Incubate for 5 min at 28 °C, 20 min at 42 °C, 40 min at 50 °C, 5 min at 80 °C before cooling down to 4 °C in the thermocycler.
10. Add 0.5 μl glycoblue and 10 μl 3 M sodium acetate (pH 5.5), mix the solution and add 250 μl ice-cold ethanol. Mix again and precipitate for 1 h to overnight at −20 °C.

3.8.2 Gel Purification of cDNAs

1. Spin for 30 min at 4 °C at 16,000 × *g*. Remove the supernatant and wash with 80% ethanol.
2. Spin for 5 min at 4 °C at 16,000 × *g*. Remove ethanol and spin briefly again to completely remove ethanol.
3. Resuspend the pellet in 6 μl H_2O.
4. Add 6 μl of 2× TBE-Urea loading buffer.
5. Heat the samples to 80 °C for 3 min and load them directly on a 5% TBE-Urea Gel. Also load a low molecular weight ladder at the far right side of the gel, for the subsequent size fractionated isolation of the cDNAs. While loading, always leave a free lane between the marker and the samples and between the samples

Fig. 3 (continued) for the preparative PCR. The expected sizes for the three fractions are 150–165 nt (low cut), 165–200 nt (medium cut), and 200–350 nt (high cut), respectively (including a total of 116 nt of linkers). In all three lanes with 30 cycle PCRs, higher migrating bands of approximately twice the expected size are visible. They correspond to cDNA concatemers that form when there have been too many PCR cycles. The free primers are visible at the bottom of the gel

themselves to have space to cut the gel. Run the gel for 40 min at 180 V until the dark blue marker reaches the lower edge.

6. Cut the marker lane from the gel and stain it for 5 min with SYBR Gold Nucleic Acid Gel Stain. Take a picture of the marker next to a fluorescent ruler. Use this ruler to measure and mark the regions of the gel that need to be cut. Cut three regions at 70–85 nt (low cut), 85–120 nt (medium cut) and 120–300 nt (high cut). Note that the 3′ linker and the RT primer together add 48 nt to the original size of the RNA fragment.
7. Transfer the gel pieces in 1.5 ml reaction tubes, add 400 μl of TE and crush the gel pieces with a pestle.
8. Incubate for 1 h at 37 °C in the thermomixer at 1100 rpm.
9. Prepare Costar Spin X columns with two 1 cm glass prefilters and transfer the solution into the columns. Spin the columns 1 min at 16,000 × *g* at room temperature (*see* **Note 21**).
10. Remove the column and add 0.5 μl glycoblue and 40 μl 3 M sodium acetate (pH 5.5), mix the solution and add 1 ml ice-cold ethanol. Mix again and precipitate for 1 h to overnight at −20 °C.

3.9 Circularization and Relinearization of the cDNAs

1. Spin for 30 min at 4 °C at 16,000 × *g*. Remove the supernatant and wash with 80% ethanol.
2. Spin for 5 min at 4 °C at 16,000 × *g*. Remove ethanol and spin briefly again to completely remove ethanol.
3. Resuspend the pellet in 6.5 μl H_2O.
4. Transfer the cDNA solution into tubes fitting the thermocycler.
5. Add 1.5 μl circularization mix:
   - 0.8 μl 10× CircLigase II buffer.
   - 0.4 μl 50 mM $MnCl_2$.
   - 0.3 μl CircLigase II.
6. Mix briefly and spin down to collect the liquid at the bottom of the tube.
7. Incubate for 1 h at 60 °C in the thermocycler.
8. Add 30 μl of oligo annealing mix:
   - 25 μl H_2O.
   - 4 μl 10× FastDigest buffer.
   - 1 μl 10 μM oligoAN.
9. Mix briefly and spin down to collect the liquid at the bottom of the tube.

10. Anneal the oligonucleotide by incubating at 95 °C for 5 min and decreasing the temperature by 1 °C every min until reaching 25 °C in the thermocycler (*see* **Note 22**).
11. Add 2 μl of FastDigest *Bam*HI and incubate at 37 °C for 30 min.
12. Add 0.5 μl glycoblue and 10 μl 3 M sodium acetate (pH 5.5), mix the solution and add 250 μl ice-cold ethanol. Mix again and precipitate for 1 h to overnight at −20 °C.

3.10 PCR and Preparation for Deep Sequencing (See Note 23)

1. Spin for 30 min at 4 °C at 16,000 × *g*. Remove the supernatant and wash with 80% ethanol.
2. Spin for 5 min at 4 °C at 16,000 × *g*. Remove ethanol and spin briefly again to completely remove ethanol.
3. Resuspend the pellet in 20 μl H_2O.
4. Use 3 μl of the cDNAs for an analytic PCR (*see* **Note 24**). Add 12 μl of PCR mix:
   - 0.3 μl of oP3/oP5 10 μM oligo mix.
   - 4.2 μl H_2O.
   - 7.5 μl Accuprime Supermix 1.
5. Run the following cycler program: 94 °C for 2 min [94 °C for 15 s, 65 °C for 30 s, 68 °C for (15/17/25) s] 25–30 cycles (*see* **Note 23**).
6. Take, for example, 5 μl of the PCR after cycle 25 and cycle 27, and let the remaining 5 μl until cycle 30. PCR should not run for more than 35 cycles.
7. Mix the PCR product aliquots with 3.75 μl of TBE loading buffer and run on a 5% TBE-urea gel at 180 V for 40 min. Stain the gel with SYBR Gold Nucleic Acid Gel Stain and analyze on a gel imager (Fig. 3b).
8. For the preparative PCR use the remaining 17 μl of cDNA and add 19 μl of PCR mix:
   - 0.72 μl of oP3/oP5 10 μM oligo mix.
   - 0.28 μl H_2O.
   - 18 μl Accuprime Supermix 1.
9. Run the following cycler program: 94 °C for 2 min [94 °C for 15 s, 65 °C for 30 s, 68 °C for (15/17/25) s] for the number of cycles determined using the analytic PCR.
10. Analyze the quality of the cDNA libraries with an Agilent analyzer (*see* **Note 25**).
11. Mix 10 nM of PCR of each of the low, medium and high fractions of the same experiment.

12. For multiplexed sequencing, mix 10 nM of each of different libraries, which carry different experiment specific barcodes (*see* **Note 26**).
13. Send a volume of the mixed libraries for high-throughput sequencing, according to the requirement of the sequencing facility (*see* **Notes 27** and **28** for considerations about sequencing data).

4 Notes

1. The water used throughout the experiment should be RNase- and DNase-free water. A good way to keep the water clean is to open a fresh bottle of water and aliquot it into 2 ml reaction tubes. A fresh aliquot should be used every day.
2. Flies start laying eggs 2 days after hatching and have their peak of egg laying 3–5 days after hatching. A good laying frequency is maintained up to a week. To obtain optimal egg collections, it is essential that the individuals are well fed from larval stages. Females are mixed with males in a 3:1 ratio. Before starting egg collections, the flies should be kept in the laying cage for at least 1 day, to get used to the cage. During this time, the red fly food dishes with fresh baker's yeast should be changed regularly. Flies can hold back eggs in their genital tract; in order to obtain well-timed egg layings, the first two to three 1 h-egg layings should be discarded, each day. To increase the number of embryos per collection, set up several laying cages simultaneously. We used females expressing GFP-Aub crossed with wild-type (w^{1118}) males [21]. Those females were obtained by crossing females bearing the *UASp-GFP-Aub* transgene with males bearing the *nos-Gal4:VP16* transgene [25]. The no-antigen negative control was performed using wild-type (w^{1118}) embryos, without transgenes.
3. The strength of the bleach solution can vary. Observe the process closely and do not keep dechorionated embryos in the bleach solution. Dechorionated embryos have a shiny appearance and float on the surface when the sieve is carefully moved up and down in the bleach solution.
4. During UV cross-linking the embryos should always be at the same distance from the bulbs to keep the same conditions in different experiments.
5. A crucial control is to perform the iCLIP on embryos where the UV cross-link is omitted. The embryos should be treated exactly as the cross-linked embryos, except for the actual cross-link.

6. Fill roughly 100 μl embryos in one reaction tube, as the experiment is carried out with 100 μl embryos. Let the embryos sink to the bottom of the 1.5 ml reaction tube, then place a P1000 tip on the edge of the embryos and swiftly remove all the PBS.
7. The lysis buffer, high salt buffer, and PNK wash should be freshly prepared for each experiment.
8. In addition to the no-cross-link negative control, two other negative controls are introduced. The no-antibody control is performed using nonimmune rabbit serum; the no-antigen control is performed with the same GFP polyclonal rabbit antibody as in the experiment, but with wild-type embryos (w^{1118} embryos, no GFP).
9. The low RNase treatment should limit the size of the RNA fragments to approximately 50–300 nt. The RNase dilution for the low RNase condition has to be established for each protein of interest. The low RNase condition should produce a smear starting from the size of the protein on the autoradiograph. The high RNase treatment should produce a sharper band of the size of the protein.
10. The truncated T4 Rnl2 is unable to adenylate the 5′-end of the substrate and needs a preadenylated 5′ DNA fragment for the ligation reaction. We use a preadenylated linker. This prevents ligation of unspecific fragments to the RNA. In addition, the truncated T4 Rnl2 was reported to be more efficient for ligating linkers to piRNAs; piRNAs have a 2′-O-methylated 3′-end which makes the ligation with the full length T4 Rnl2 less efficient [26, 27].
11. The Dynabeads should remain in suspension during the enzymatic reactions that are performed "on beads." We perform these reactions in a thermomixer at 1000 rpm.
12. The transfer of RNPs is less efficient than the transfer of the corresponding protein alone in our hands. We obtain better results with a slower transfer overnight. The transfer was performed with ice pads in the tank, or at 4 °C when performed overnight.
13. Use a fluorescent sticker to accurately match the autoradiograph to the membrane. Precise matching of autoradiograph and membrane is mandatory to isolate the RNPs of interest.
14. In our hands, the 30 min- and 1 h-exposures gave strong signals on the autoradiograph. Expose overnight if you need to pause the experiment. No radioactive signals should be observed in the no-cross-link, the no-antibody, and the no-antigen controls. These controls are mandatory to verify that the observed radioactive labeled RNPs are specific.

15. There should be a more or less sharp band slightly above the size of the protein, in the high RNase condition. The low RNase condition should produce a smear starting at the size of the band in the high RNase condition. Seventy nucleotide of RNA have the molecular weight of approximately 20 kDa [28]. In order to isolate RNA fragments with a length of 40–300 nt (including the linker) cut out a region starting at the band in the high RNase condition, and up to 80 kDa heavier.
16. You can verify with a Geiger counter that the elution of the RNA from the membrane was successful. There should be no, or only a very low radioactive signal left on the membrane pieces after the elution.
17. The protein digestion with Proteinase K leaves a residual peptide at the cross-linked nt. This peptide acts as a hindrance for the reverse transcriptase, which causes reverse transcription to frequently stop at the nt just before the cross-linked nt. The iCLIP uses this information. The cross-linked RNA fragments are linked to a single linker at their 3′-ends. Through circularization of these molecules and cleavage of the linker after the reverse transcription, cDNAs with 3′ and 5′ linkers are generated. Most of these cDNAs stop at the nt before the cross-link site, and therefore carry the information of the protein interaction site on the RNA.
18. Washing the RNA pellet with lower ethanol concentrations (i.e., 70%) can lead to the solubilization and loss of small RNA fragments (~20 nt).
19. The RNA pellet is very small and the glycoblue helps to visualize the pellet and follow its solubilization. Do not overdry the pellet as this will make it difficult to dissolve. Completely remove ethanol with a P10 tip and dry briefly at room temperature if necessary.
20. Each RT primer is tagged with (a) a random barcode that will identify PCR duplicates in the subsequent bioinformatic analyses, and (b) an experiment specific barcode, to allow the multiplexing of different libraries for sequencing. The experiment specific barcode also allows contamination of the library with previously made libraries to be detected (*see* **Note 23**). The experiment specific barcode has a strong impact on the reverse transcription efficiency. Therefore, it is useful to test new RT primers [28].
21. Place the filters on top of the column with a pair of forceps and push them down carefully with the large end of a P10 tip.
22. This slow decrease in temperature is optimal for annealing the oligonucleotide to the circularized cDNA.

23. The PCR is performed separately for the three cDNA fractions (low, medium and high cut), such that optimal PCR conditions can be used for each size fraction. The PCR elongation times for the low, medium and high cut fractions are 15, 17, and 25 s, respectively. Because cDNA libraries are PCR amplified, to prevent contaminations of future iCLIP experiments it is essential that the PCR reactions are set up in a specific PCR room, and that all post-PCR steps are performed in a different working area than the pre-PCR steps. Contaminations can occur with the tiniest amounts of cDNAs as all libraries are amplified with the same PCR primer pair. One way to monitor contamination is to use an RT primer with a new experiment specific barcode for each experiment, or if reusing a previously used RT primer, doing so only after several experiments.
24. The PCR should amplify the cDNAs to a concentration allowing sequencing, but should not overamplify the libraries, since this might reduce their complexity. An overamplified library can contain cDNA concatemers which arise through annealing of identical adapter sequences during the PCR (Fig. 3b). To determine the right number of PCR cycles, an analytic PCR is performed before the preparative PCR. A 15 μl PCR reaction is set up with 3 μl cDNA; two times 5 μl aliquots are removed after a certain number of PCR cycles (e.g., 20 and 22 cycles), and the rest of the reaction remains until the end of the programmed PCR. The PCR aliquots are analyzed on a gel and the cycle number where the band appears first is determined. The result can be different for the low, medium or high cut fractions. The preparative PCR is then set up according to the determined cycle number. The fact that a higher cDNA concentration is used for the preparative PCR is taken into account: one PCR cycle is subtracted for a threefold increase in cDNA concentration in the preparative PCR. The PCR fragments should be 150–165, 165–200, and 200–350 nt, for the low, medium and high cuts, respectively (Fig. 3b).
25. The profile from the Agilent analyzer gives information about the size distribution and concentration of DNA in the libraries. In case the preparative PCR shows too many additional bands with the Agilent analyzer (e.g., primer dimers), the PCR fragment can be gel purified. The concentration can also be determined using RT-qPCR when a library of known concentration is available as a standard. RT-qPCR has the advantage that only DNA molecules with the correct adapters on both ends will be measured.
26. The experiment specific barcode are the first nt to be read during the sequencing. At this phase of the sequencing, important steps take place, such as cluster detection and cluster focusing, which are essential to set up the sequencing. When

different libraries are mixed for multiplexed sequencing, the experiment specific barcodes can introduce a substantial bias in sequence composition. This bias can prevent properly establishing the sequencing parameters, which in turn will affect or prevent the sequencing itself. This point should be considered when choosing experiment specific barcodes to be mixed.

27. After sequencing the reads are sorted according to their experiment specific barcode. Adapter sequences are trimmed. PCR duplicates are eliminated by merging reads with the same start and stop positions on the same strand and the same random barcode.

28. Two types of RNA fragments are expected from Aub iCLIP: piRNAs or fragments of piRNAs, and fragments of longer RNAs interacting with Aub. Because iCLIP is based on the truncation of cDNAs at the cross-link site, the recovery of complete piRNAs is not expected. However, a proportion of complete piRNAs are present in Aub iCLIPs, suggesting a cross-link at the 5′-phosphate of piRNAs. These piRNAs have the characteristics of this class of small RNAs: a bias for a start with a Uridine, a length of 24–27 nt, and their genomic origin from the piRNA clusters. The second category of RNA fragments from longer RNAs bound by Aub is expected to originate from piRNA precursors, transposable element mRNAs and cellular mRNAs. Specific bioinformatic pipelines should be developed to analyze each of these classes of RNA. The reproducibility of cross-link sites within iCLIP experiments, as well as between independent iCLIPs (biological replicates) is an important parameter to consider for identifying significant cross-link sites over background [21].

Acknowledgments

Work in the Simonelig lab is supported by the UMR9002 CNRS-Université de Montpellier, ANR (ANR-15- CE12-0019-01), FRM (Equipe FRM 2013 DEQ20130326534), and AFM-Telethon (N° 17110). B.B. was supported by D.F.G. and J.D. by Fondation ARC.

References

1. Weick EM, Miska EA (2014) piRNAs: from biogenesis to function. Development 141:3458–3471
2. De Renzis S, Elemento O, Tavazoie S, Wieschaus EF (2007) Unmasking activation of the zygotic genome using chromosomal deletions in the Drosophila embryo. PLoS Biol 5:e117
3. Rouget C, Papin C, Boureux A, Meunier AC, Franco B et al (2010) Maternal mRNA deadenylation and decay by the piRNA pathway in the early Drosophila embryo. Nature 467:1128–1132

4. Goh WS, Falciatori I, Tam OH, Burgess R, Meikar O et al (2015) piRNA-directed cleavage of meiotic transcripts regulates spermatogenesis. Genes Dev 29:1032–1044
5. Gou LT, Dai P, Yang JH, Xue Y, Hu YP et al (2014) Pachytene piRNAs instruct massive mRNA elimination during late spermiogenesis. Cell Res 24:680–700
6. Watanabe T, Cheng EC, Zhong M, Lin H (2015) Retrotransposons and pseudogenes regulate mRNAs and lncRNAs via the piRNA pathway in the germline. Genome Res 25 (3):368–380
7. Zhang P, Kang JY, Gou LT, Wang J, Xue Y et al (2015) MIWI and piRNA-mediated cleavage of messenger RNAs in mouse testes. Cell Res 25(2):193–207
8. Kiuchi T, Koga H, Kawamoto M, Shoji K, Sakai H et al (2014) A single female-specific piRNA is the primary determiner of sex in the silkworm. Nature 509:633–636
9. Licatalosi DD, Mele A, Fak JJ, Ule J, Kayikci M et al (2008) HITS-CLIP yields genome-wide insights into brain alternative RNA processing. Nature 456:464–469
10. Chi SW, Zang JB, Mele A, Darnell RB (2009) Argonaute HITS-CLIP decodes microRNA-mRNA interaction maps. Nature 460:479–486
11. Leung AK, Young AG, Bhutkar A, Zheng GX, Bosson AD et al (2011) Genome-wide identification of Ago2 binding sites from mouse embryonic stem cells with and without mature microRNAs. Nat Struct Mol Biol 18:237–244
12. Zisoulis DG, Lovci MT, Wilbert ML, Hutt KR, Liang TY et al (2010) Comprehensive discovery of endogenous argonaute binding sites in Caenorhabditis elegans. Nat Struct Mol Biol 17:173–179
13. Broughton JP, Pasquinelli AE (2016) A tale of two sequences: microRNA-target chimeric reads. Genet Sel Evol 48:31
14. Konig J, Zarnack K, Rot G, Curk T, Kayikci M et al (2010) iCLIP reveals the function of hnRNP particles in splicing at individual nucleotide resolution. Nat Struct Mol Biol 17:909–915
15. Sugimoto Y, Konig J, Hussain S, Zupan B, Curk T et al (2012) Analysis of CLIP and iCLIP methods for nucleotide-resolution studies of protein-RNA interactions. Genome Biol 13:R67
16. Bosson AD, Zamudio JR, Sharp PA (2014) Endogenous miRNA and target concentrations determine susceptibility to potential ceRNA competition. Mol Cell 56:347–359
17. Broughton JP, Lovci MT, Huang JL, Yeo GW, Pasquinelli AE (2016) Pairing beyond the seed supports microRNA targeting specificity. Mol Cell 64:320–333
18. Broughton JP, Pasquinelli AE (2013) Identifying argonaute binding sites in Caenorhabditis elegans using iCLIP. Methods 63:119–125
19. Vourekas A, Zheng Q, Alexiou P, Maragkakis M, Kirino Y et al (2012) Mili and Miwi target RNA repertoire reveals piRNA biogenesis and function of Miwi in spermiogenesis. Nat Struct Mol Biol 19:773–781
20. Sytnikova YA, Rahman R, Chirn GW, Clark JP, Lau NC (2014) Transposable element dynamics and PIWI regulation impacts lncRNA and gene expression diversity in Drosophila ovarian cell cultures. Genome Res 24(12):1977–1990
21. Barckmann B, Pierson S, Dufourt J, Papin C, Armenise C et al (2015) Aubergine iCLIP reveals piRNA-dependent decay of mRNAs involved in germ cell development in the early embryo. Cell Rep 12:1205–1216
22. Vourekas A, Alexiou P, Vrettos N, Maragkakis M, Mourelatos Z (2016) Sequence-dependent but not sequence-specific piRNA adhesion traps mRNAs to the germ plasm. Nature 531:390–394
23. Konig J, Zarnack K, Rot G, Curk T, Kayikci M et al (2011) iCLIP—transcriptome-wide mapping of protein-RNA interactions with individual nucleotide resolution. J Vis Exp, JoVE
24. Walter J, Biggin MD (1997) Measurement of in vivo DNA binding by sequence-specific transcription factors using UV cross-linking. Methods 11:215–224
25. Harris AN, Macdonald PM (2001) Aubergine encodes a Drosophila polar granule component required for pole cell formation and related to eIF2C. Development 128:2823–2832
26. Grentzinger T, Armenise C, Brun C, Mugat B, Serrano V et al (2012) piRNA-mediated transgenerational inheritance of an acquired trait. Genome Res 22:1877–1888
27. Munafo DB, Robb GB (2010) Optimization of enzymatic reaction conditions for generating representative pools of cDNA from small RNA. RNA 16:2537–2552
28. Huppertz I, Attig J, D'Ambrogio A, Easton LE, Sibley CR et al (2014) iCLIP: protein-RNA interactions at nucleotide resolution. Methods 65:274–287

Chapter 8

Integration of ENCODE RNAseq and eCLIP Data Sets

Jorge Boucas

Abstract

During the last decade, the study of mRNA decay has largely benefited from an increasing number of high-throughput assays that emerged from developments in next generation sequencing (NGS) technologies as well as mass spectrometry. While assay-specific data analysis is often reported and software made available many researchers struggle with the overwhelming challenge of integrating data from diverse assays, different sources, and of different formats.

We here use *Python*, *R*, and *bash* to analyze and integrate RNAseq and eCLIP data publicly available from ENCODE. Annotation is performed with biomart, motif analysis with MEME and finally a functional enrichment analysis using DAVID. This analysis is centered on KHSRP eCLIP data from K562 cell as well as RNAseq data from KHSRP knockdown and respective mock controls.

Key words ENCODE, RNAseq, eCLIP, Bioinformatics, mRNA decay, Python, R, bash

1 Introduction

In 1984 Miller and colleagues reported for the first time that modification of the 3′ untranslated region (UTR) of *c-fos* allows efficient transformation of NIH-3T3 cells [1]. Later, Treisman and colleagues showed that depletion of an AU-rich element (ARE) on the 3′ of *c-fos* significantly increases the stability of *c-fos* mRNA [2]. As in the case of *c-fos*, many other genes have been shown to contain AREs as *cis* regulatory elements in their 3′UTR that promote deadenylation and degradation of the mRNA body [3, 4]. ARE binding proteins (AUBPs), are *trans* regulatory elements that target AREs and can subsequently modulate the degradation process [4]. microRNAs (miRNAs), ~22-nucleotide RNAs, are another group of *trans* regulatory elements that target mRNAs [5–7]. Once assembled as a part of the RNA-induced silencing complex (RISC) a miRNA can define the specificity of this complex

Electronic supplementary material: The online version of this article (https://doi.org/10.1007/978-1-4939-7540-2_8) contains supplementary material, which is available to authorized users.

Shireen R. Lamandé (ed.), *mRNA Decay: Methods and Protocols*, Methods in Molecular Biology, vol. 1720, https://doi.org/10.1007/978-1-4939-7540-2_8,

and therefore interfere with the stability of specific target transcripts. Presnyak and colleagues [8] recently performed a genome-wide RNA decay assay and found that both translation and mRNA decay are fine-tuned by the balance between optimal and nonoptimal codons. During the last decade, the study of mRNA decay has largely benefited from an increasing number of high-throughput assays that emerged from developments in next generation sequencing (NGS) technologies as well as mass spectrometry, e.g., RNA-sequencing (RNAseq), for quantifying transcripts levels; photoactivatable ribonucleotide-enhanced crosslinking and immunoprecipitation (PAR-CLIP [9]) and related—HITS-CLIP [9], iCLIP [10, 11]—for identifying binding sites of RNA-binding proteins as well as miRNA-mRNA interactions in ribonucleoprotein (RNPs) complexes; mRNA-RBP interactome capture [12, 13] for identifying all proteins interacting with mRNA at a defined time point. Along with this, large amounts of data have been generated, deposited in open repositories, databases created, data standards implemented, and software developed. While assay-specific data analysis is often reported and software made available many researchers struggle with the overwhelming challenge of integrating data from diverse assays, different sources, and of different formats. The increasing number of software options has become in many cases overwhelming as well (*see* **Note 1**).

Having made its way into the world of big data, in biological research the concept of data wrangling—conversion or mapping data from one form into another—for convenient use of the data is still poorly understood and integrated. As with most other processes in research it is important to first understand the input as well as the desired output before deciding on a way to approach the problem. Thus, one should become familiar with the most common data formats associated with NGS (e.g., fasta, fastq, gtf, bed, SAM, vcf—*see* **Note 2**) and understand the value of reference repositories e.g., UCSC [14, 15], Ensembl [16]. Ensembl offers gold standards for assembled genomes, i.e., reference DNA, RNA, and protein sequences, as well as gene sets and variation annotations. Accessible through a web interface and in a programmatic fashion (API [17] and database dumps) Ensembl offers multispecies data as well as single species data. Similar and complementary information is provided by the UCSC Genome Browser. It is here important to notice the use of different coordinate systems (*see* **Note 3**). Most of Ensembl data can be further accessed through BioMart [18], a querying tool available both through a web interface and in a programmatic fashion. Upon new releases all data is transferred to archive.ensembl.org and stored for at least 3 years allowing researchers to seamlessly continue work initiated on previous releases.

The Encyclopedia of DNA Elements [19] (ENCODE) collects a comprehensive list of functional genomic elements and offers the respective data in raw, and treated formats. Promoted by the National Cancer Institute and the National Human Genome Research Institute, The Cancer Genome Atlas (TCGA) offers descriptive data from tumor tissue of more than 11,000 patients. Other standard repositories are currently used to deposit NGS data, e.g., European Nucleotide Archive [20, 21] (ENA; also, Sequence Read Archive, SRA); Gene Expression Omnibus (GEO) [22]. With a focus on miRNA research several databases and prediction tools have emerged as well, e.g., miRBase [23] and TargetScan [24]. The same holds true for AREs (AU-Rich Element-containing mRNA Database [25], ARED), UTRs (The Atlas of UTR Regulatory Activity [26], AURA) and general posttranscriptional regulation and/or motif analysis, e.g., doRiNA [27], starBase [28], MEME [29], and HOMER [30].

Founded in 1998, the Gene Ontology [31, 32] (GO) project has an extensive collection of descriptions of gene products and is one of the most commonly used resources for functional annotation of genes. While tools like QuickGO and AmiGO allow easy exploration of GO terms, others, like DAVID [33, 34] and TopGO [35] allow enrichment analysis of provided datasets.

We will here use *Python*, *R*, and *bash* to analyze and integrate RNAseq and eCLIP data from ENCODE. Annotation will be performed with biomart, motif analysis with MEME and finally a functional enrichment analysis using DAVID. This analysis will be centered on KHSRP eCLIP data from K562 cell as well as RNAseq data from KHSRP knockdown and respective mock controls.

2 Materials

2.1 Software

1. All the software for this pipeline is preinstalled in a docker image. You will need to have *Docker*—https://www.docker.com—and *git*—https://www.atlassian.com/git/tutorials/install-git—installed. The following instructions will work for UNIX systems. Users running other operating systems will also be able to use the analysis pipeline provided *Docker* and *git* are installed.
2. Once *Docker* and *git* are installed, start by cloning the repository into your home directory. In *bash*: *cd ~/ && git clone https://github.com/mpg-age-bioinformatics/ENCODE_RNAseq_eCLIP*

   Then build the Docker image:
   cd ~/ENCODE_RNAseq_eCLIP && sudo docker build -t encode_rnaseq_eclip.
   With this completed all software is installed and ready to use from the *Docker* image.

3. Users who wish to run the *Jupyter notebook* without *Docker* should install *Conda*—http://conda.pydata.org—and make sure they have installed *Jupyter*—http://jupyter.org. Afterward, one should be able follow the installation in the *Dockerfile*—https://github.com/mpg-age-bioinformatics/ENCODE_RNAseq_eCLIP/docker/Dockerfile.

2.2 Usage

The analysis pipeline was tested on a *Docker* image running with 8 GB of memory and four CPUS.

1. The container can be started in *bash* with:

   *sudo docker run -d -p 8888:8888 *
   *-e GRANT_SUDO = yes -e NB_UID = 1000 --user root *
   *-e PASSWORD = "YOURPASS" -e USE_HTTPS = yes *
   *-v ~/ENCODE_RNAseq_eCLIP/notebooks:/home/jovyan/work/notebooks *
   *-v ~/ENCODE_RNAseq_eCLIP/results:/home/jovyan/work/results *
   *--name encode_rnaseq_eclip *
   -i -t encode_rnaseq_eclip

   Instead of *"YOURPASS"* type in a password of your choice.
2. Then, on your web browser connect to https://localhost:8888. You will be requested to accept the self-signed certificate and to give in the password you used in *"YOURPASS"*.
3. In the Notebook Dashboard navigate to find the notebook: clicking on its name will open it in a new browser tab.
4. Click on the menu Help → User Interface Tour for an overview of the Jupyter Notebook App user interface. You can run the notebook document step-by-step (one cell a time) by pressing shift + enter. You can run the whole notebook in a single step by clicking on the menu Cell → Run All. To restart the kernel (i.e., the computational engine), click on the menu Kernel → Restart. This can be useful to start over a computation from scratch (variables are deleted, open files are closed, etc.. . .).
5. Stop the container with *docker stop encode_rnaseq_eclip* and restart it with *docker start encode_rnaseq_eclip*.

2.3 Prior-Knowledge

Preferentially, users will have a minimal introduction to programming. Given the easy syntax of the languages used in this pipeline—mostly *Python*—users without a previous introduction to programming will also be able to follow this analysis step-by-step. Readers should be able to understand variables, that everything typed after a # is a comment and not part of the code, and that everything between quotes is a string, e.g., *"This is a string"*. An introduction to both *Python* and *bash* can be gain at codeacademy.com. An *R*

```
def MY_FUNCTION(x):
    """
    Documentation
    """
    value=operation
    return value

def DOUBLE(x):
    """
    This functon returns the
    double of a given number
    """"
    y=x*2
    return y
```

Fig. 1 Defining functions in Python

course is also available at datacamp.com. In Python, functions are defined as shown in Fig. 1.

Using functions can be extremely practical once the code needs to be applied to more than one input group. We will mostly define functions for our operations as we will tend to reuse them, e.g., gene data and isoform data.

2.4 Data

ENCODE [19] Data Coordination Center (DCC) or GEO accession numbers of the datasets (DCC accession: ENCSR037HRJ; GEO accession: GSE30567).

3 Methods

All code and results matching the different steps can be seen on the annexed html (*see* Electronic Supplemental Material on link.springer.com) as well as on the notebooks of the respective *git* repository—https://github.com/mpg-age-bioinformatics/ENCODE_RNAseq_eCLIP.

3.1 Download Raw Data, Reference Genome, and Reference Annotation

Having identified KHSRP as a molecule of interest in ENCODE we start by downloading the files for the knockdown of KHSRP in the human cell line K562. The association graph in ENCODE (https://www.encodeproject.org/experiments/ENCSR561CBC/) shows two replicates for which transcript and gene quantifications are already available. We collect these files for further analysis. It is important to make sure that all downloaded data which is already processed has the same reference genome and annotation. In this case, we use the GRCh38 v24 reference. Equally, the controls shown in the summary section are downloaded from ENCODE.

Reference fasta and GTF files for different releases are available at GENCODE.

3.2 Differential Gene Expression

In ENCODE, by selecting the different analysis steps on the association graph we are shown which analysis pipeline has been used for processing the data—"Long RNA-seq RSEM quantification step for paired-end pipeline." RSEM [36] is a software package for estimating gene and isoform expression levels from RNA-Seq data. RSEM is distributed with the downstream tool EBseq [37] which can be used for differential gene expression analysis. In *bash*, we can check how a program works by doing for example *rsem-run-ebseq --help*.

We follow the instructions and generate differential gene expression tables from the data downloaded from ENCODE.

3.3 Parsing Annotation File

The annotation.gtf file contains valuable annotation information. It is therefore practical to have it in an easy to use format like a DataFrame. As this step can be quite consuming we execute it first and store the output in a file.

3.4 Import Required Packages

In python we can import packages and attribute them a new (short) name, e.g., *import pandas as pd*.

We then use functions from the specific package with *pd.FunctionName(arguments)*.

If we want to get help for the respective function we use the *help* function, e.g., *help(pd.FunctionName)*. This can be quite practical when invoking functions from a package that might exist with the same name in another package.

3.5 Explore Differential Gene Expression Results

1. We start by reading in the output of EBseq into a python Pandas DataFrame. We plot the distribution of mean counts (expression) for each sample and plot the expression of each gene in the knockdown sample in function of its expression in the mock control. Significantly changed genes tend to accumulate toward the higher levels of expression as they are also easier to detect and quantify (Fig. 2).
2. To identify genes strongly changed in relation to others with the same expression level we plot MA like plots (Fig. 2) of the *log2(fold change)* of each gene in function of its *normalized intensities*, i.e., *log10(sqrt(expression in condition 1 * expression in condition 2))*.
3. We divide the genes through bins depending on their normalized intensities, identify the corresponding 0.5 *log2(fold change)* percentile for each bin, and fit a polynomial curve.

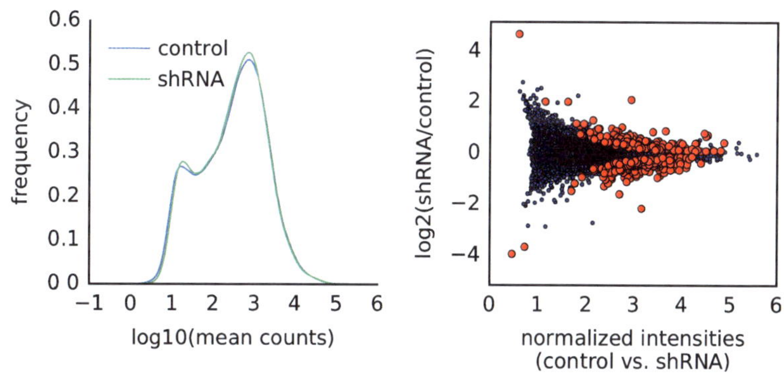

Fig. 2 Distribution of gene expression values (*left*) and respective MA plot (*right*; *red*—significantly changed genes)

4. To identify genes of interest we mark genes out of the 0.5 percentile that are also significantly changed with red. It is important to collect these genes in a list as later on we will want to analyze them further.

3.6 Read eCLIP Peaks

1. After reading the bed narrow peak files from ENCODE we transform the datafames into a pybedtools bedtool object to be able to merge both replicates into one single bedtool.
2. We discard peaks where both replicates show a *p*-value above 0.05.
3. We then annotate each peak with the overlapping exons, transcripts, and genes for each respective gene. Additionally, we count the number of times each feature (i.e., exon, transcript, and gene) is targeted and normalize the values of the narrow peak files for each feature.

3.7 Merge RNAseq and eCLIP Data

1. For merging RNAseq and eCLIP data we start by replotting MA plots highlighting target genes identified by eCLIP (Fig. 3, upper). As highly expressed genes interacting with RBPs will be easier to detect in eCLIP than lower expressed genes the distribution of the counts for KHSRP targets shows the respective bias toward targets with a higher level of expression (Fig. 3, lower). We plot as well an MA like plot where the log2(fold change) for each transcript is plotted in function of the number of respective KHSRP binding sites.

3.8 Biotype and Gene Ontology Annotation

Gene ontology is one of the best examples of the value of ontologies and the richest resource of annotations on gene function and cellular localization. For annotating our tables with gene ontology terms as well as gene biotype we will use the biomaRt package for *R* and Ensembl's biomart service. It is here important to realize that Ensembl releases a new version of its biomart database with each genome release and that older database releases are transferred to

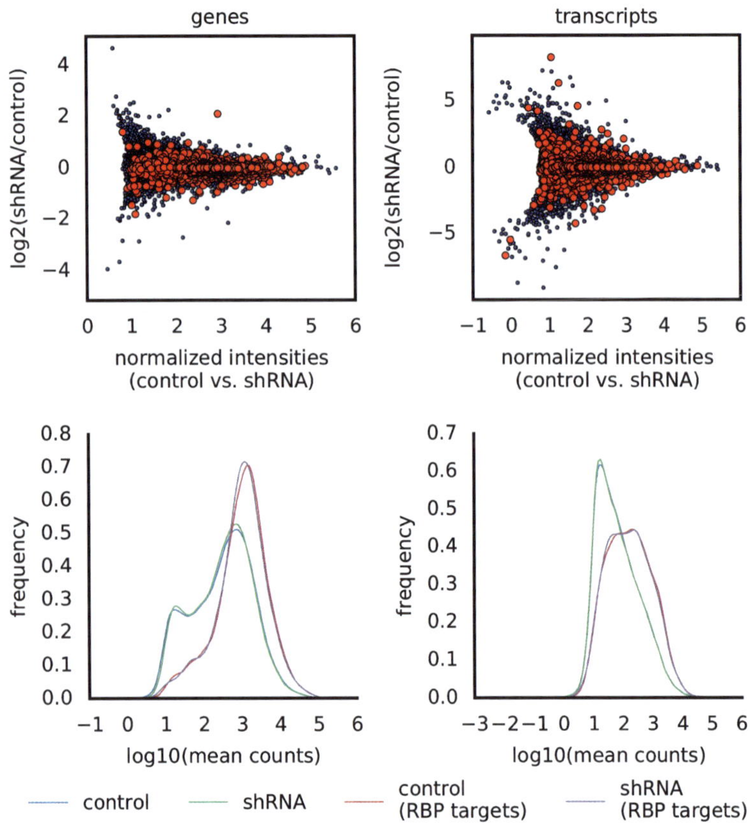

Fig. 3 MA plots (*upper*, *red*—KHSRP eCLIP targets) and distribution of gene expression for whole transcriptome and eCLIP-identified KHSRP targets

archive.ensembl.org. In our case, the header of our downloaded GTF clearly indicates we require a release matching "Ensembl 83"—dec2015.archive.ensembl.org.

3.9 Clustering and Heatmaps

1. We start by preparing a DataFrame with log10(expression) values. We concatenate gene_id and gene_name for each gene as we will later on use this to label the rows (i.e., genes) in our heatmap.
2. To identify the number of clusters that best describe our results we plot the 'within groups sum of squares' versus the number of clusters. This is often more an art than a science but once the 'within groups sum of squares' does not change with the increase in number of clusters this defines the optimal number of clusters to separate our data into.

3.10 Writing Report Tables

Having different gene sets recovered from our MA plots as well as annotated tables we save our DataFrames into excel files using a separate sheet for each set.

3.11 Enrichment Analysis with DAVID

The Database for Annotation, Visualization, and Integrated Discovery—DAVID—is a powerful tool for identifying enriched GO terms in a gene set. Furthermore DAVID also allows the analysis of terms from other databases as for example: KEGG, PFAM, and OMIM. DAVID has a user friendly web frontend as well as a practical API. We here use the DAVID API to programmatically analyze all our gene sets. We create a small loop that reads each file generated in Subheading 3.10 and for each sheet inside submits a query to DAVID, retrieves the result and saves it into a new excel report file.

3.12 Generating MOTIFs with MEME

1. With the plots above demonstrating a bias in eCLIP toward identifying KHSRP targets with high levels of expression we use the peaks identified by eCLIP to generate a KHSRP target motif consensus. We will later run this motif through all human transcripts to identify target transcripts in an unbiased way. We here use meme-cip, part of the MEME suite. For eCLIP, meme-chip requires 100 base sequences centered on the peak center.
2. Having generated the BED file with the regions of interest for defining the target peak we use *bedtools getfasta* to extract the respective sequences from the reference genome fasta file in a stranded specific fashion. As we deal with RNA and not DNA we convert all "T"s to "U"s with *sed*. Otherwise, meme-chip will start looking for motifs on the reverse strand as well.
3. Before defining a motif as useful for identifying targets on a transcriptome wide scale we define a motif as valid if the number of sites it encounters during meme-chip is equal or greater than 5% of the number of given sites used for the generation of the motif.

3.13 Identifying Target Transcripts Using MEME

We start by extracting the fasta sequences of each transcript using *gffread*, the annotation.gft file as well as the genome.fa file. As before, we convert all Ts to U. Finally, we run *fimo* to scan the transcriptome for the presence of our selected motifs.

3.14 Merging MEME Identified Targets and DGE

As before, we visualize the final result of MEME combined with our differential gene expression results (Fig. 4). We here expect to lose the bias toward genes with higher levels of expression. We annotate our table of MEME identified targets, save the respective report tables, and use DAVID for enrichment analysis.

3.15 Calculating Distance between Target Sites and Stop Codons

1. KHSRP is a key mediator of mRNA decay through the interaction with AU-rich-elements in target mRNAs [38]. Given this, we expect KHSRP target sites to be located in the 3′UTR of respective target mRNAs. We therefore calculated the distance between each target site and the stop codon of each respective target gene. We

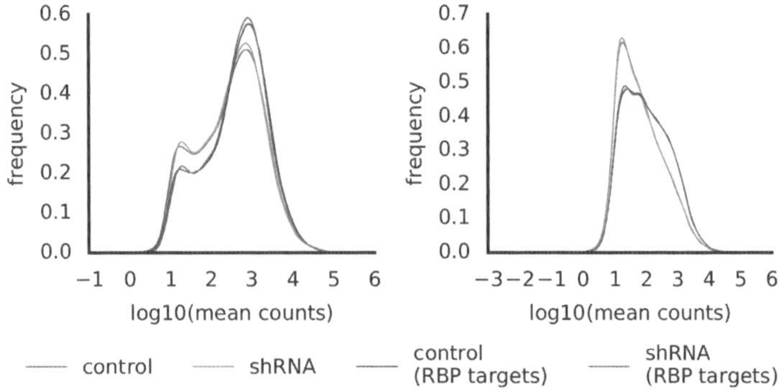

Fig. 4 Distribution of expression levels for genes (*left*) and transcripts (*right*) in whole transcriptome and computationally identified (FIMO) KHSRP targets (i.e., RBP targets)

start by creating a dictionary where for each transcript all exon positions are present, e.g., dic = {"ESNT00001":[1001,1002, 1003,..0.6023,6024,6025], "ESNT00001":[3301,3302,3303, ..0.5042,5043,5044]}. It is important to notice that for transcripts encoded in the "-" strand, the positions of each exon needs to be reverse sorted.

2. Having created this dictionary we can use it to identify the position of each genomic coordinate on the transcript. For example, Using the dictionary above we can see that position 1003 of "ESNT00001" is the third position in the transcript. Having translated all genomic coordinates of target sites and stop codons to transcript coordinates we plot the distribution of the distances between both (Fig. 5).

3.16 Filtering Terms of Interest from Enrichment Tables

We have previously shown that Khsrp is regulated in response to genotoxic stress [39]. We therefore searched for any "Biological process" enriched term in all DAVID "Biological process" outputs which contain at least one of the strings "cell cycle," "apoptosis," "cell death," "cell division," or "proliferation."

4 Notes

1. Links

Ensembl: http://www.ensembl.org
UCSC Genome Browser: http://genome.ucsc.edu
Ensembl web interface for data download: http://www.ensembl.org/info/data/ftp/index.html
Ensembl BioMart API: http://www.ensembl.org/info/data/biomart/index.html
Ensembl FTP server: ftp.ensembl.org
BioMart: http://www.biomart.org

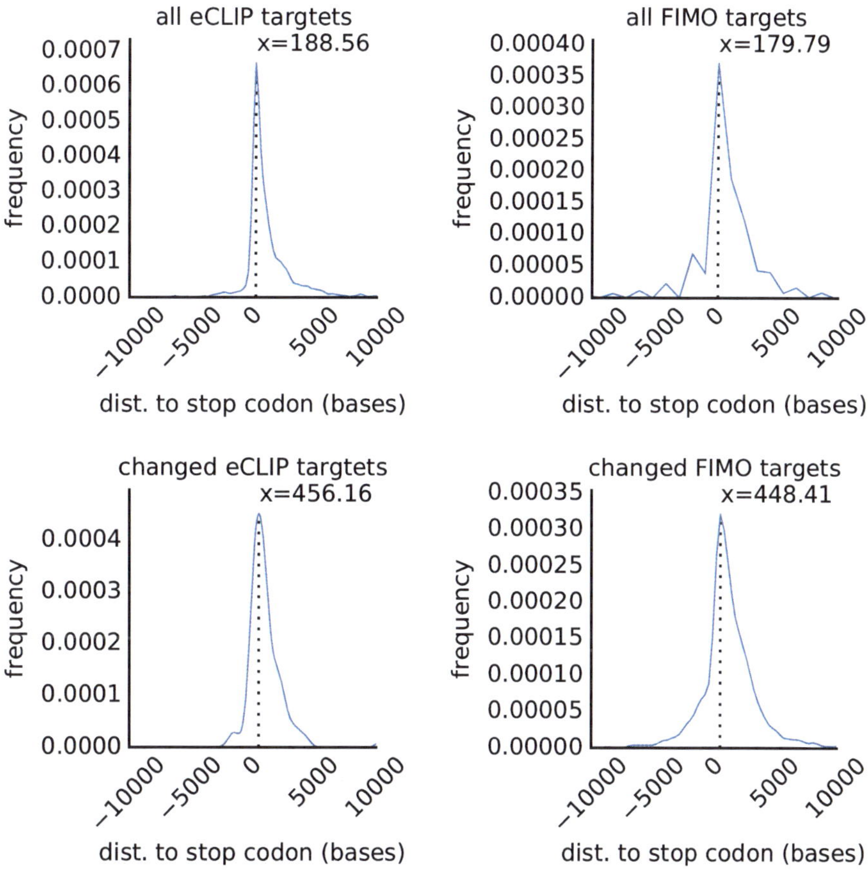

Fig. 5 Distance between KHSRP binding sites and respective stop codons for eCLIP-identified targets (*left*) and FIMO identified targets (*right*). Significantly changed transcripts we analyzed separately (*lower*)

Ensembl BioMart: http://www.ensembl.org/biomart/martview/

ENCODE, The Encyclopedia of DNA Elements: https://www.encodeproject.org

GENCODE, gene sets used by the ENCODE and the 1000 Genomes consortium: https://www.gencodegenes.org

TCGA, The Cancer Genome Atlas: http://cancergenome.nih.gov

ENA, European Nucleotide Archive: http://www.ebi.ac.uk/ena

SRA, Sequence Read Archive: https://www.ncbi.nlm.nih.gov/sra

GEO, Gene Expression Omnibus: https://www.ncbi.nlm.nih.gov/geo/

miRBase, microRNA database: http://www.mirbase.org

TargetScan, prediction of biological targets of miRNAs: http://www.targetscan.org

ARED, AU-Rich Element-containing mRNA Database: http://brp.kfshrc.edu.sa/ARED/

AURA, Atlas of UTR Regulatory Activity: http://aura.science.unitn.it

doRiNA, database of RNA interactions in post-transcriptional regulation: http://dorina.mdc-berlin.de

starBase, decoding Pan-Cancer and Interaction Networks: http://starbase.sysu.edu.cn

MEME, motif-based sequence analysis tools: http://meme-suite.org

HOMER, motif discovery and next generation sequencing analysis: http://homer.salk.edu

GO, Gene Ontology Consortium: http://geneontology.org

QuickGo, a fast browser for Gene Ontology terms and annotations: http://www.ebi.ac.uk/QuickGO/

AmiGO, searching and browsing the Gene Ontology database: http://amigo.geneontology.org/

DAVID, Database for Annotation, Visualization and Integrated Discovery: https://david.ncifcrf.gov

TopGO, topology-based Gene Ontology scoring: http://topgo.bioinf.mpi-inf.mpg.de

codeacademy, interactive learn how to program: https://www.codeacademy.com

biomaRt: an R package for the access of major bioinformatics databases: https://bioconductor.org/packages/release/bioc/vignettes/biomaRt/inst/doc/biomaRt.html

archive.ensembl.org, ENSEMBL's archived releases: http://www.ensembl.org/info/website/archives/index.html

2. File formats

fasta

As defined in https://blast.ncbi.nlm.nih.gov/Blast.cgi?CMD=Web&PAGE_TYPE=BlastDocs&DOC_TYPE=BlastHelp

A sequence in FASTA format begins with a single-line description, followed by lines of sequence data. The description line (defline) is distinguished from the sequence data by a greater-than (">") symbol at the beginning. It is recommended that all lines of text be shorter than 80 characters in length. Blank lines are not allowed in the middle of FASTA input.

fastq

As defined in http://maq.sourceforge.net/fastq.shtml

FASTQ format stores sequences and Phred qualities in a single file. It is concise and compact. FASTQ is first widely used in the Sanger Institute and therefore we usually take the Sanger specification and the standard FASTQ format, or simply FASTQ format. Although Solexa/Illumina read file looks pretty much like FASTQ, they are different in that the qualities are scaled differently. In the quality string, if you can see a character with its ASCII code higher than 90, probably your file is in the Solexa/Illumina format.

<fastq>	:=	<block>+
<block>	:=	@<seqname>\n<seq>\n+[<seqname>]\n<qual>\n
<seqname>	:=	[A-Za-z0-9_.:-]+
<seq>	:=	[A-Za-z\n\.~]+
<qual>	:=	[!-~\n]+

gtf

As defined in http://www.ensembl.org/info/website/upload/gff.html

Fields **must** be tab-separated. Also, all but the final field in each feature line must contain a value; "empty" columns should be denoted with a '.'.

seqname	Name of the chromosome or scaffold; chromosome names can be given with or without the 'chr' prefix
source	Name of the program that generated this feature, or the data source (database or project name)
feature	Feature type name, e.g., Gene, Variation, Similarity
start	Start position of the feature, with sequence numbering starting at 1
end	End position of the feature, with sequence numbering starting at 1
score	A floating point value
strand	defined as + (forward) or – (reverse)
frame	One of "0," "1," or "2," "0" indicates that the first base of the feature is the first base of a codon, "1" that the second base is the first base of a codon, and so on
attribute	A semicolon-separated list of tag-value pairs, providing additional information about each feature

Sequence Alignment/Map (SAM)

These are the plain text output produced by alignment tools after mapping of reads (*see* fastq format) to a reference genome (*see* fasta format). Two compressed, binary versions exist: bam and cram.

As defined in http://www.htslib.org/doc/samtools.html:

1	QNAME	Query template/pair NAME
2	FLAG	bitwise FLAG
3	RNAME	Reference sequence NAME
4	POS	1-based leftmost POSition/coordinate of clipped sequence

(continued)

5	MAPQ	MAPping Quality (Phred-scaled)
6	CIGAR	extended CIGAR string
7	MRNM	Mate Reference sequence NaMe ('=' if same as RNAME)
8	MPOS	1-based Mate POSistion
9	TLEN	inferred Template LENgth (insert size)
10	SEQ	query SEQuence on the same strand as the reference
11	QUAL	query QUALity (ASCII-33 gives the Phred base quality)
12+	OPT	variable OPTional fields in the format TAG:VTYPE: VALUE

Each bit in the FLAG field is defined as:

0x0001	p	The read is paired in sequencing
0x0002	P	The read is mapped in a proper pair
0x0004	u	The query sequence itself is unmapped
0x0008	U	The mate is unmapped
0x0010	r	Strand of the query (1 for reverse)
0x0020	R	Strand of the mate
0x0040	1	The read is the first read in a pair
0x0080	2	The read is the second read in a pair
0x0100	s	The alignment is not primary
0x0200	f	The read fails platform/vendor quality checks
0x0400	d	The read is either a PCR or an optical duplicate
0x0800	S	The alignment is supplementary

where the second column gives the string representation of the FLAG field.

bed

As defined in https://genome.ucsc.edu/FAQ/FAQformat.html#format1:

BED (Browser Extensible Data) lines have three required tab separated fields (1–3) and nine (4–12) additional optional fields.

Fields:

1	**chrom**	The name of the chromosome (e.g., chr3, chrY, chr2_random) or scaffold (e.g., scaffold10671)
2	**chromStart**	The starting position of the feature in the chromosome or scaffold. The first base in a chromosome is numbered 0

(continued)

3	**chromEnd**	The ending position of the feature in the chromosome or scaffold. The *chromEnd* base is not included in the display of the feature. For example, the first 100 bases of a chromosome are defined as *chromStart=0*, *chromEnd=100*, and span the bases numbered 0–99
4	**name**	Defines the name of the BED line. This label is displayed to the left of the BED line in the Genome Browser window when the track is open to full display mode or directly to the left of the item in pack mode
5	**score**	A score between 0 and 1000. If the track line *useScore* attribute is set to 1 for this annotation data set, the *score* value will determine the level of gray in which this feature is displayed (higher numbers = darker gray)
6	**strand**	Defines the strand—either "+" or "−"
7	**thickStart**	The starting position at which the feature is drawn thickly (for example, the start codon in gene displays). When there is no thick part, thickStart and thickEnd are usually set to the chromStart position
8	**thickEnd**	The ending position at which the feature is drawn thickly (e.g., the stop codon in gene displays)
9	**itemRgb**	An RGB value of the form R,G,B (e.g., 255,0,0). If the track line *itemRgb* attribute is set to "On", this RBG value will determine the display color of the data contained in this BED line. NOTE: It is recommended that a simple color scheme (eight colors or less) be used with this attribute to avoid overwhelming the color resources of the Genome Browser and your Internet browser
10	**blockCount**	The number of blocks (exons) in the BED line
11	**blockSizes**	A comma-separated list of the block sizes. The number of items in this list should correspond to *blockCount*
12	**blockStarts**	A comma-separated list of block starts. All of the *blockStart* positions should be calculated relative to *chromStart*. The number of items in this list should correspond to *blockCount* In BED files with block definitions, the first *blockStart* value must be 0, so that the first block begins at *chromStart*. Similarly, the final *blockStart* position plus the final *blockSize* value must equal *chromEnd*. Blocks may not overlap

ENCODE narrowPeak

As defined in https://genome.ucsc.edu/FAQ/FAQformat.html#format12:

This format is used to provide called peaks of signal enrichment based on pooled, normalized (interpreted) data. It is a BED6+4 format.

Fields:

1	**chrom**	Name of the chromosome (or contig, scaffold, etc.)
2	**chromStart**	The starting position of the feature in the chromosome or scaffold. The first base in a chromosome is numbered 0
3	**chromEnd**	The ending position of the feature in the chromosome or scaffold. The *chromEnd* base is not included in the display of the feature. For example, the first 100 bases of a chromosome are defined as *chromStart=0, chromEnd=100*, and span the bases numbered 0–99
4	**name**	Name given to a region (preferably unique). Use '.' if no name is assigned
5	**score**	Indicates how dark the peak will be displayed in the browser (0–1000). If all scores were "0" when the data were submitted to the DCC, the DCC assigned scores 1–1000 based on signal value. Ideally the average signalValue per base spread is between 100–1000
6	**strand**	+/- to denote strand or orientation (whenever applicable). Use "." if no orientation is assigned
7	**signalValue**	Measurement of overall (usually, average) enrichment for the region
8	**qValue**	Measurement of statistical significance (−log10). Use −1 if no *p*Value is assigned
9	**qValue**	Measurement of statistical significance using false discovery rate (−log10). Use −1 if no *q*Value is assigned
10	**peak**	Point-source called for this peak; 0-based offset from chromStart. Use −1 if no point-source called

3. Others

Coordinate systems

While Ensembl uses a 1-based coordinate system, UCSC uses a 0-based coordinate system, i.e., in the 1-based coordinate system, single nucleotides, variants positions, or ranges are specified by their corresponding nucleotide numbers; in the 0-based coordinate system, single nucleotides, variants positions, or ranges are specified by the coordinates that flank them. The usage of different coordinates systems further expands to file formats, with GFF, SAM, and VCF being 1-based, and BED and BAM being 0-based [40]. Thus, when wrangling these files it is often useful to validate the performed transformations by inspecting the original sequence at stake as well as the final one. This can also be done by visual inspection of the sequence in the respective coordinates in Ensembl.

Acknowledgments

ENCODE Consortium, the labs of Brenton Graveley at UConn and Gene Yeo at UCSD for generating the KHSRP knockdown and eCLIP data respectively.

References

1. Miller AD, Curran T, Verma IM (1984) c-fos protein can induce cellular transformation: a novel mechanism of activation of a cellular oncogene. Cell 36(1):51–60
2. Wilson T, Treisman R (1988) Removal of poly (A) and consequent degradation of c-fos mRNA facilitated by 3′ AU-rich sequences. Nature 336(6197):396–399. https://doi.org/10.1038/336396a0
3. Shyu AB, Belasco JG, Greenberg ME (1991) Two distinct destabilizing elements in the c-fos message trigger deadenylation as a first step in rapid mRNA decay. Genes Dev 5(2):221–231
4. Xu N, Chen CY, Shyu AB (1997) Modulation of the fate of cytoplasmic mRNA by AU-rich elements: key sequence features controlling mRNA deadenylation and decay. Mol Cell Biol 17(8):4611–4621
5. Nilsen TW (2007) Mechanisms of microRNA-mediated gene regulation in animal cells. Trends Genet 23(5):243–249. https://doi.org/10.1016/j.tig.2007.02.011
6. Filipowicz W, Bhattacharyya SN, Sonenberg N (2008) Mechanisms of post-transcriptional regulation by microRNAs: are the answers in sight? Nat Rev Genet 9(2):102–114. https://doi.org/10.1038/nrg2290
7. Guo H, Ingolia NT, Weissman JS, Bartel DP (2010) Mammalian microRNAs predominantly act to decrease target mRNA levels. Nature 466(7308):835–840. https://doi.org/10.1038/nature09267
8. Presnyak V, Alhusaini N, Chen YH, Martin S, Morris N, Kline N, Olson S, Weinberg D, Baker KE, Graveley BR, Coller J (2015) Codon optimality is a major determinant of mRNA stability. Cell 160(6):1111–1124. https://doi.org/10.1016/j.cell.2015.02.029
9. Danan C, Manickavel S, Hafner M (2016) PAR-CLIP: a method for transcriptome-wide identification of RNA binding protein interaction sites. Methods Mol Biol 1358:153–173. https://doi.org/10.1007/978-1-4939-3067-8_10
10. Hafner M, Landthaler M, Burger L, Khorshid M, Hausser J, Berninger P, Rothballer A, Ascano M Jr, Jungkamp AC, Munschauer M, Ulrich A, Wardle GS, Dewell S, Zavolan M, Tuschl T (2010) Transcriptome-wide identification of RNA-binding protein and microRNA target sites by PAR-CLIP. Cell 141(1):129–141. https://doi.org/10.1016/j.cell.2010.03.009
11. Konig J, Zarnack K, Rot G, Curk T, Kayikci M, Zupan B, Turner DJ, Luscombe NM, Ule J (2010) iCLIP reveals the function of hnRNP particles in splicing at individual nucleotide resolution. Nat Struct Mol Biol 17 (7):909–915. https://doi.org/10.1038/nsmb.1838
12. Baltz AG, Munschauer M, Schwanhausser B, Vasile A, Murakawa Y, Schueler M, Youngs N, Penfold-Brown D, Drew K, Milek M, Wyler E, Bonneau R, Selbach M, Dieterich C, Landthaler M (2012) The mRNA-bound proteome and its global occupancy profile on protein-coding transcripts. Mol Cell 46(5):674–690. https://doi.org/10.1016/j.molcel.2012.05.021
13. Castello A, Fischer B, Eichelbaum K, Horos R, Beckmann BM, Strein C, Davey NE, Humphreys DT, Preiss T, Steinmetz LM, Krijgsveld J, Hentze MW (2012) Insights into RNA biology from an atlas of mammalian mRNA-binding proteins. Cell 149 (6):1393–1406. https://doi.org/10.1016/j.cell.2012.04.031
14. Kent WJ, Sugnet CW, Furey TS, Roskin KM, Pringle TH, Zahler AM, Haussler D (2002) The human genome browser at UCSC. Genome Res 12(6):996–1006. https://doi.org/10.1101/gr.229102. Article published online before print in May 2002
15. Rosenbloom KR, Armstrong J, Barber GP, Casper J, Clawson H, Diekhans M, Dreszer TR, Fujita PA, Guruvadoo L, Haeussler M, Harte RA, Heitner S, Hickey G, Hinrichs AS, Hubley R, Karolchik D, Learned K, Lee BT, Li CH, Miga KH, Nguyen N, Paten B, Raney BJ, Smit AF, Speir ML, Zweig AS, Haussler D, Kuhn RM, Kent WJ (2015) The UCSC Genome Browser database: 2015 update. Nucleic Acids Res 43(Database issue): D670–D681. https://doi.org/10.1093/nar/gku1177

16. Yates A, Akanni W, Amode MR, Barrell D, Billis K, Carvalho-Silva D, Cummins C, Clapham P, Fitzgerald S, Gil L, Giron CG, Gordon L, Hourlier T, Hunt SE, Janacek SH, Johnson N, Juettemann T, Keenan S, Lavidas I, Martin FJ, Maurel T, McLaren W, Murphy DN, Nag R, Nuhn M, Parker A, Patricio M, Pignatelli M, Rahtz M, Riat HS, Sheppard D, Taylor K, Thormann A, Vullo A, Wilder SP, Zadissa A, Birney E, Harrow J, Muffato M, Perry E, Ruffier M, Spudich G, Trevanion SJ, Cunningham F, Aken BL, Zerbino DR, Flicek P (2016) Ensembl 2016. Nucleic Acids Res 44 (D1):D710–D716. https://doi.org/10.1093/nar/gkv1157

17. Yates A, Beal K, Keenan S, McLaren W, Pignatelli M, Ritchie GR, Ruffier M, Taylor K, Vullo A, Flicek P (2015) The Ensembl REST API: Ensembl data for any language. Bioinformatics 31(1):143–145. https://doi.org/10.1093/bioinformatics/btu613

18. Smedley D, Haider S, Durinck S, Pandini L, Provero P, Allen J, Arnaiz O, Awedh MH, Baldock R, Barbiera G, Bardou P, Beck T, Blake A, Bonierbale M, Brookes AJ, Bucci G, Buetti I, Burge S, Cabau C, Carlson JW, Chelala C, Chrysostomou C, Cittaro D, Collin O, Cordova R, Cutts RJ, Dassi E, Di Genova A, Djari A, Esposito A, Estrella H, Eyras E, Fernandez-Banet J, Forbes S, Free RC, Fujisawa T, Gadaleta E, Garcia-Manteiga JM, Goodstein D, Gray K, Guerra-Assuncao JA, Haggarty B, Han DJ, Han BW, Harris T, Harshbarger J, Hastings RK, Hayes RD, Hoede C, Hu S, Hu ZL, Hutchins L, Kan Z, Kawaji H, Keliet A, Kerhornou A, Kim S, Kinsella R, Klopp C, Kong L, Lawson D, Lazarevic D, Lee JH, Letellier T, Li CY, Lio P, Liu CJ, Luo J, Maass A, Mariette J, Maurel T, Merella S, Mohamed AM, Moreews F, Nabihoudine I, Ndegwa N, Noirot C, Perez-Llamas C, Primig M, Quattrone A, Quesneville H, Rambaldi D, Reecy J, Riba M, Rosanoff S, Saddiq AA, Salas E, Sallou O, Shepherd R, Simon R, Sperling L, Spooner W, Staines DM, Steinbach D, Stone K, Stupka E, Teague JW, Dayem Ullah AZ, Wang J, Ware D, Wong-Erasmus M, Youens-Clark K, Zadissa A, Zhang SJ, Kasprzyk A (2015) The BioMart community portal: an innovative alternative to large, centralized data repositories. Nucleic Acids Res 43(W1):W589–W598. https://doi.org/10.1093/nar/gkv350

19. Consortium EP (2012) An integrated encyclopedia of DNA elements in the human genome. Nature 489(7414):57–74. https://doi.org/10.1038/nature11247

20. Leinonen R, Sugawara H, Shumway M, International Nucleotide Sequence Database C (2011) The sequence read archive. Nucleic Acids Res 39(Database issue):D19–D21. https://doi.org/10.1093/nar/gkq1019

21. Kodama Y, Shumway M, Leinonen R, International Nucleotide Sequence Database C (2012) The Sequence Read Archive: explosive growth of sequencing data. Nucleic Acids Res 40 (Database issue):D54–D56. https://doi.org/10.1093/nar/gkr854

22. Barrett T, Wilhite SE, Ledoux P, Evangelista C, Kim IF, Tomashevsky M, Marshall KA, Phillippy KH, Sherman PM, Holko M, Yefanov A, Lee H, Zhang N, Robertson CL, Serova N, Davis S, Soboleva A (2013) NCBI GEO: archive for functional genomics data sets—update. Nucleic Acids Res 41(Database issue): D991–D995. https://doi.org/10.1093/nar/gks1193

23. Kozomara A, Griffiths-Jones S (2014) miRBase: annotating high confidence microRNAs using deep sequencing data. Nucleic Acids Res 42(Database issue):D68–D73. https://doi.org/10.1093/nar/gkt1181

24. Agarwal V, Bell GW, Nam JW, Bartel DP (2015) Predicting effective microRNA target sites in mammalian mRNAs. elife 4. https://doi.org/10.7554/eLife.05005

25. Halees AS, El-Badrawi R, Khabar KS (2008) ARED Organism: expansion of ARED reveals AU-rich element cluster variations between human and mouse. Nucleic Acids Res 36(Database issue):D137–D140. https://doi.org/10.1093/nar/gkm959

26. Dassi E, Re A, Leo S, Tebaldi T, Pasini L, Peroni D, Quattrone A (2014) AURA 2: empowering discovery of post-transcriptional networks. Translation (Austin) 2(1):e27738. https://doi.org/10.4161/trla.27738

27. Blin K, Dieterich C, Wurmus R, Rajewsky N, Landthaler M, Akalin A (2015) DoRiNA 2.0—upgrading the doRiNA database of RNA interactions in post-transcriptional regulation. Nucleic Acids Res 43(Database issue): D160–D167. https://doi.org/10.1093/nar/gku1180

28. Li JH, Liu S, Zhou H, Qu LH, Yang JH (2014) starBase v2.0: decoding miRNA-ceRNA, miRNA-ncRNA and protein-RNA interaction networks from large-scale CLIP-Seq data. Nucleic Acids Res 42(Database issue):D92–D97. https://doi.org/10.1093/nar/gkt1248

29. Bailey TL, Boden M, Buske FA, Frith M, Grant CE, Clementi L, Ren J, Li WW, Noble WS (2009) MEME SUITE: tools for motif discovery and searching. Nucleic Acids Res 37(Web

Server issue):W202–W208. https://doi.org/10.1093/nar/gkp335
30. Heinz S, Benner C, Spann N, Bertolino E, Lin YC, Laslo P, Cheng JX, Murre C, Singh H, Glass CK (2010) Simple combinations of lineage-determining transcription factors prime cis-regulatory elements required for macrophage and B cell identities. Mol Cell 38 (4):576–589. https://doi.org/10.1016/j.molcel.2010.05.004
31. Ashburner M, Ball CA, Blake JA, Botstein D, Butler H, Cherry JM, Davis AP, Dolinski K, Dwight SS, Eppig JT, Harris MA, Hill DP, Issel-Tarver L, Kasarskis A, Lewis S, Matese JC, Richardson JE, Ringwald M, Rubin GM, Sherlock G (2000) Gene ontology: tool for the unification of biology. The Gene Ontology Consortium. Nat Genet 25(1):25–29. https://doi.org/10.1038/75556
32. Gene Ontology C (2015) Gene Ontology Consortium: going forward. Nucleic Acids Res 43(Database issue):D1049–D1056. https://doi.org/10.1093/nar/gku1179
33. Huang d W, Sherman BT, Lempicki RA (2009) Bioinformatics enrichment tools: paths toward the comprehensive functional analysis of large gene lists. Nucleic Acids Res 37(1):1–13. https://doi.org/10.1093/nar/gkn923
34. Huang d W, Sherman BT, Lempicki RA (2009) Systematic and integrative analysis of large gene lists using DAVID bioinformatics resources. Nat Protoc 4(1):44–57. https://doi.org/10.1038/nprot.2008.211
35. Alexa A, Rahnenfuhrer R (2016) topGO: enrichment analysis for gene ontology. R package version 2280
36. Li B, Dewey CN (2011) RSEM: accurate transcript quantification from RNA-Seq data with or without a reference genome. BMC Bioinformatics 12:323. https://doi.org/10.1186/1471-2105-12-323
37. Leng N, Dawson JA, Thomson JA, Ruotti V, Rissman AI, Smits BM, Haag JD, Gould MN, Stewart RM, Kendziorski C (2013) EBSeq: an empirical Bayes hierarchical model for inference in RNA-seq experiments. Bioinformatics 29(8):1035–1043. https://doi.org/10.1093/bioinformatics/btt087
38. Trabucchi M, Briata P, Garcia-Mayoral M, Haase AD, Filipowicz W, Ramos A, Gherzi R, Rosenfeld MG (2009) The RNA-binding protein KSRP promotes the biogenesis of a subset of microRNAs. Nature 459 (7249):1010–1014. https://doi.org/10.1038/nature08025
39. Boucas J, Fritz C, Schmitt A, Riabinska A, Thelen L, Peifer M, Leeser U, Nuernberg P, Altmueller J, Gaestel M, Dieterich C, Reinhardt HC (2015) Label-free protein-RNA interactome analysis identifies Khsrp signaling downstream of the p38/Mk2 kinase complex as a critical modulator of cell cycle progression. PLoS One 10(5):e0125745. https://doi.org/10.1371/journal.pone.0125745
40. Griffith O (2013) Tutorial: cheat sheet for one-based vs zero-based coordinate systems. https://www.biostars.org/p/84686/

Chapter 9

Identifying miRNA Targets Using AGO-RIPseq

Rebecca Petri and Johan Jakobsson

Abstract

microRNAs (miRNA) are small, noncoding RNAs that bind to messenger RNAs (mRNAs) and regulate their activity. They are, therefore, important posttranscriptional regulators. In recent years it has become clear that miRNAs regulate large genetic networks, rather than single genes, and that one gene can be targeted by several miRNAs. To understand the role of miRNAs in cells or tissues, it is therefore important to analyze the targetome of miRNAs. Here, we present a technique called Argonaute-RNA Immunoprecipitation (AGO-RIP) which takes advantages of the fact that miRNAs and their targets are directly bound by the Argonaute protein family. With this approach quantitative, genome-wide analysis of miRNA targets is possible. In this chapter we describe the RIP-methodology and provide advice for RNA sequencing and bioinformatic analyses.

Key words microRNAs, Targetome, Argonaute, Immunoprecipitation, Small and total RNA sequencing

1 Introduction

miRNAs are small, 21–23 nucleotides, single-stranded and endogenously expressed RNAs that have emerged as important posttranscriptional regulators. miRNAs bind messenger RNAs (mRNAs) and regulate their translational inhibition or degradation [1, 2].

miRNAs are transcribed from genomic DNA by RNA Polymerase II or III to generate a long, primary miRNA transcript (pri-miRNA) which typically contains a stem loop structure. The pri-miRNA is processed by a multiprotein complex, Microprocessor, consisting of DGCR8 and the RNaseIII enzyme Drosha, to a 50–70 nucleotides long precursor miRNA (pre-miRNA). Subsequently, the pre-miRNA is exported into the cytoplasm by the nuclear export protein Exportin 5, where the stem loop structure of the pre-miRNA is recognized and further cleaved by the enzymes Dicer and TAR RNA-binding protein 2 (TRBP) to form mature miRNA duplexes. One strand of this duplex is typically stabilized and bound by Argonaute proteins (AGO) to form the RNA-induced silencing complex (RISC), while the other strand is

Shireen R. Lamandé (ed.), *mRNA Decay: Methods and Protocols*, Methods in Molecular Biology, vol. 1720, https://doi.org/10.1007/978-1-4939-7540-2_9, © Springer Science+Business Media, LLC 2018

degraded. The AGO bound miRNA then guides the RISC to its respective mRNA targets. The binding sites are commonly in the mRNA 3′UTR. miRNA–target interaction is predominantly directed by the miRNA's "seed sequence" which ranges from 6 to 8 bases in the 5′ end of the miRNA [1, 3, 4].

miRNAs have been implicated in a plethora of biological functions, highlighting their importance in posttranscriptional regulation. In recent years, an increasing number of studies suggested that miRNAs regulate large genetic networks, rather than single genes [5, 6]. This is supported by the fact that mRNA and protein levels of genes are only marginally affected by miRNA targeting [5, 6]. Moreover, one mRNA can be targeted by several miRNAs and targets of a specific miRNA are often functionally connected in distinct pathways.

To understand the role of miRNAs it is important to identify miRNA target genes in distinct tissues and cells. Several algorithms were developed to predict miRNA target sites in the 3′UTR of genes based on the miRNA seed sequence. Although these algorithms have improved, they are not completely overlapping and still lead to false-positive findings, making experimental approaches to investigate the miRNA targetome necessary [1, 7, 8].

In recent years, we and others have shown that biochemical isolation of AGO proteins using techniques such as RIP-seq, HITS-CLIP and CLASH enables the isolation of miRNAs and their targets from different cells and tissues [9–14], taking advantage of their direct binding [15].

In this chapter, we describe an AGO-RIP approach followed by RNA extraction to analyze miRNAs and their targets in cells and tissues without cross-linking to quantitatively assess transcript abundance in the RISC. We moreover give advice on small and total RNA sequencing and the bioinformatic analyses (Fig. 1).

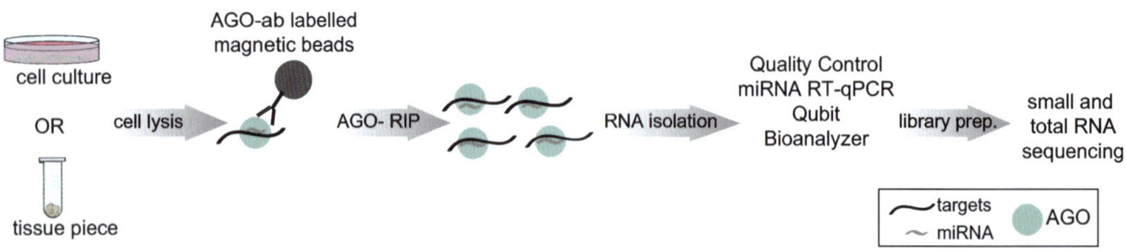

Fig. 1 Schematics of the AGO-RIP procedure. Cultured cells or freshly harvested tissue can be used as starting material for the AGO-RIP. Upon cell/tissue lysis, the AGO protein is immunoprecipitated using magnetic beads labeled with AGO antibodies and RNA is extracted. The quality of the RIP is controlled by determining the enrichment of miRNAs, and by the measurement of the concentration and size distribution of the RNA samples. After the quality control, cDNA libraries are prepared and small or total RNA sequencing followed by bioinformatic analyses are conducted

2 Materials

Prepare all buffers in nuclease-free water immediately before starting the AGO-RIP procedure. Always use gloves and avoid RNase contamination by wiping benches and equipment with RNaseZAP or equivalent products. Always use filter pipette tips. It is also advisable to use reagents only for this technique to avoid contaminations (*see* **Note 1**).

2.1 Buffers

1. PBST: 1× PBS (pH 7.4), 0.02% Tween.
2. Lysis buffer: 10 mM HEPES (pH 7.3), 5 mM $MgCl_2$, 100 mM KCl, 0.5% NP40, EDTA-free protease inhibitor cocktail; immediately before use add 0.5 mM DTT.
3. Complete lysis buffer: 10 mM HEPES (pH 7.3), 5 mM $MgCl_2$, 100 mM KCl, 0.5% NP40, EDTA-free protease inhibitor cocktail; immediately before use add 0.5 mM DTT, 1 mM PMSF, and 400 U/ml RNase inhibitor.
4. Low salt NT2 buffer: 50 mM Tris–HCl (pH 7.5), 1 mM $MgCl_2$, 150 mM NaCl, 0.5% NP40, EDTA-free protease inhibitor cocktail; immediately before use add 0.5 mM DTT, 1 mM PMSF, and 100 U/ml RNase Inhibitor.
5. High salt NT2 buffer: 50 mM Tris–HCl (pH 7.5), 1 mM $MgCl_2$, 600 mM NaCl, 0.5% NP40; immediately before use add 0.5 mM DTT, EDTA-free protease inhibitor cocktail, 1 mM PMSF, and 100 U/ml RNase Inhibitor.

2.2 Cell/Tissue Lysis

1. TissueLyser LT (Qiagen).
2. Stainless steel bead (Qiagen).

2.3 RNA Immunoprecipitation

1. Magnetic stand for 1.5 ml tubes.
2. Dynabeads® Protein G.
3. Antibodies against endogenous Argonaute proteins (e.g., Sigma 2E12-1C9 for AGO2).
4. Tube rotator.

2.4 RNA Extraction

1. miRNA and total RNA extraction kit (e.g., miRNeasy micro kit, Qiagen).
2. DNase.

2.5 Quality Control

1. Qubit (Qubit microRNA Assay Kit and Qubit RNA HS Assay Kit).
2. Bioanalyzer (e.g., Agilent smallRNA analyses kit and Agilent RNA 6000 Pico Kit).

3. cDNA synthesis kit for miRNAs (e.g., Exiqon, Universal cDNA Synthesis Kit II).
4. PCR tubes.
5. SYBR Green Master mix.

2.6 cDNA Library Preparation

1. cDNA library preparation for total RNA (e.g., Nugen, Ovation® RNA-Seq System V2 and Ovation® Ultralow System).
2. cDNA library kit for small RNA sequencing (e.g., NEB, NEB-Next® Small RNA Library Prep Set for Illumina®).

2.7 Bioinformatic Analyses

1. Cutadapt [16].
2. FastQC (http://www.bioinformatics.babraham.ac.uk/projects/fastqc/).
3. STAR aligner [17].
4. Subread package FeatureCounts [18].
5. Annotation files from for example miRbase [19], Ensembl [20], UCSC Genome Browser [21], and NCBI Refseq [22].

3 Methods

The following protocol describes an AGO-RIP protocol followed by RNA extraction of miRNAs and their targets bound by endogenously expressed AGO protein in cells or tissues. The protocol includes the preparation of beads and buffers, the AGO-RIP, RNA extraction and advice on RNA sequencing and the bioinformatic analyses.

3.1 Preparation of Beads and Buffers for AGO-RIP

1. Prepare PBST according to the material section and keep it on ice.
2. In the meantime adjust the bottle with the Dynabeads to room temperature. Vortex or shake the bottle so that the beads deattach from the walls of the bottle.
3. Pipette 200 μl Dynabeads per sample into a clean 1.5 ml tube (*see* **Note 2**).
4. Put the tube with the magnetic beads into a magnetic stand and wait for 1 min until all beads are attached to the magnet and the solution is clear. Discard the supernatant.
5. Carefully add 1000 μl of PBST. For this, place the pipette tip directly above the beads. Carefully rinse them from the tube walls without touching them directly.
6. Repeat the washing step with PBST.
7. Dilute the antibody in PBST (*see* **Note 3**).

8. Incubate the beads with the antibody for a maximum of 1 h (*see* **Notes 4** and **5**).
9. After the incubation step, wash the beads twice with PBST and twice with lysis buffer. Pipette the same amount of beads in lysis buffer into 1.5 ml tubes, one tube per sample. Keep the beads in lysis buffer on ice until further use.

3.2 Preparation of Cell/Tissue Lysates

1. Prepare the lysis buffer and complete lysis buffer according to the material section and keep on ice.
2. Harvest a cell pellet or a tissue piece of interest. Try to be as quick as possible and work clean when dissecting tissue to avoid RNase contaminations. Rinse the tissue with ice-cold PBS to remove blood remains and immediately add to a 2 ml tube with 500–1000 μl of complete lysis buffer and a stainless steel bead. If harvesting a cell pellet, pellet cells in a 2 ml tube and add 500–1000 μl complete lysis buffer and a steel bead (*see* **Notes 6–8**).
3. Samples should be immediately lysed using the TissueLyser LT for 2 minutes and 30 Hz. Put the adaptor on ice prior to use (*see* **Notes 9** and **10**).
4. After lysis, transfer the lysates to clean 1.5 ml tubes and keep them on ice.

3.3 RNA Immunoprecipitation

1. Clear the lysates by centrifuging for 15 min at 16,200 × *g*, 4 °C.
2. Transfer the supernatant into a new 1.5 ml tube (*see* **Note 11**).
3. Save 1/10 of the lysate as INPUT sample (= total RNA control) and keep on ice.
4. Put the AGO-antibody labeled magnetic beads into the magnetic stand. Wait 1 min until the liquid is clear. Remove the supernatant.
5. Add the remaining cell lysate to the respective beads and gently resuspend.
6. Incubate the lysates 24 h at 4 °C with end-over-end rotation (*see* **Note 12**).

3.4 RNA Extraction from INPUT Sample

1. Add 700 μl QIAzol to the INPUT aliquot of the cell lysate and incubate for 10 min at room temperature.
2. Continue the RNA extraction according to the miRNeasy micro kit protocol (*see* **Note 13**).
3. After the elution of the RNA, keep the samples at −80 °C until further use.

3.5 RNA Immunoprecipitation

1. After a 24 h incubation of the lysate with the magnetic beads, prepare the low and high salt NT2 buffers directly before starting the RIP and store on ice.
2. Put the magnetic stand on ice.
3. Collect the beads by putting the tubes on the magnetic stand.
4. Wait until all the beads have attached and the liquid is clear.
5. Remove the supernatant (RNA is now bound to the beads).
6. Wash the beads carefully by rinsing them off the tube wall with 1000 μl low salt NT2 buffer (*see* **Note 14**). Resuspend several times and transfer to a new and clean 1.5 ml tube.
7. Repeat the washing step.
8. Repeat the washing step twice with high salt NT2 buffer.
9. After the last washing step remove the liquid and resuspend the beads in 700 μl Qiazol.
10. Incubate for 10 min at room temperature.
11. After 10 min of incubation, put the tubes on the magnetic stand and wait until all the beads are attached to the side of the tube.
12. Transfer the Qiazol solution (contains the RNA) to a new and clean 1.5 ml tube.
13. Continue according to the miRNeasy micro kit by Qiagen or equivalent kit and use the same steps as for the RNA extraction from INPUT samples.
14. After eluting the RNA, store the samples at −80 °C or directly proceed with the quality control to avoid freezing-thawing cycles.

3.6 Quality Control

3.6.1 Qubit and Bioanalyzer

1. To determine the concentration and size distribution of the samples use Qubit and bioanalyzer.
2. Also keep in mind that ribosomal RNA cannot be used for testing RNA integrity of RIP samples, since most of the ribosomal RNA is lost during the RIP procedure.

3.6.2 Quantitative RT-PCR

1. Run RT-qPCR to test whether the enrichment of miRNAs in the RIP using the AGO antibody has worked compared to a control antibody.
2. Prepare cDNA from miRNAs using the lowest possible concentration to save material for sequencing. Prepare according to the kit guidelines (*see* **Note 15**).
3. In brief, take the amount of your sample and add nuclease-free water up to 14 μl. Add 4 μl of master mix and finally 2 μl of enzyme. Always keep the enzyme in an ice block and add it last to the reaction.

4. Immediately after adding the enzyme, start the RT-PCR with the program suggested in the respective kit.
5. After cDNA synthesis, dilute cDNA 1:5 in nuclease-free water and run qPCR using SYBR Green Master Mix (*see* **Notes 16** and **17**). Always pipette the PCR reactions in triplicates.
6. Primers for miRNAs, which are known to be expressed and bound by AGO in the tissue/cells of interest should be used to calculate the enrichment of miRNAs. Primers for miRNAs that are known not to be active in the cells/tissue of interest can be used as control.
7. Calculate the average ct-value of triplicates. Calculate the enrichment of a miRNA by subtracting the average ct-value of this miRNA in the AGO-RIP fraction from the ct-value in the background control sample. Calculate the fold change enrichment (ΔΔCt) to validate enrichment of miRNAs in your RIP fraction.

3.7 Preparation of cDNA Sequencing Libraries and Sequencing

1. The procedures of the cDNA library preparation and the RNA sequencing are not described in detail here. The kits used for cDNA library preparation are mentioned in the material section (*see* **Note 18**).
2. Single read 1 × 50 bp sequencing and an output of around 5–10 M reads for small RNA sequencing and 15–30 M for RNA sequencing, depending on the research question, are recommended.

3.8 Bioinformatic Analysis

The bioinformatic analysis has to be designed for each experiment, depending on the design of the experiments and the research questions being asked. In general the following steps are conducted:

1. Remove 3′ adapter sequences from reads using appropriate tools (e.g., Cutadapt [16]).
2. Run quality control on your fastq files (e.g., FastQC, Babraham Bioinformatics, http://www.bioinformatics.babraham.ac.uk/projects/fastqc/). For small RNA sequencing a distinct peaks at 21–23 nucleotides should be seen after the trimming of adapter sequences.
3. Use an appropriate alignment tool (e.g., STAR aligner [17]) to align to the respective genome.
4. Quantify the reads (e.g., Subread package FeatureCounts [18]) and normalize the dataset appropriately depending on the type of experiment.
5. The small RNA sequencing data can be used to characterize which small RNAs are bound by AGO in specific cells or tissues. Annotation files can be obtained from miRbase

(http://www.mirbase.org/ [19]) for miRNA analyses and for grouping of individual miRNAs into miRNA families. The dataset can be used to show which miRNAs are expressed and/or enriched in the AGO fraction compared to INPUT total RNA controls. Enrichment of miRNAs in the RISC can be calculated by dividing the reads in RIP samples by that of INPUT samples.

6. For total RNA sequencing data, a characterization of mRNAs or noncoding RNAs bound by AGO can be conducted. Annotation files for mRNAs or noncoding RNAs can for example be obtained by Ensembl [20], NCBI RefSeq [22], or UCSC Genome Browser [21]. As for miRNAs, enrichment of target genes or noncoding RNAs in the RISC can be calculated by dividing the reads in RIP samples by that of INPUT samples.

4 Notes

1. It is recommended to use an extra pipette set for the AGO-RIP procedure to avoid contaminations.
2. The bead solution can be very viscous. Pipette small amounts so that the correct amount of beads will be used (e.g., twice 100 μl instead of 200 μl at once).
3. The optimal antibody concentration has to be tested and optimized for each antibody prior to the experiment. We usually use the antibodies in a concentration of 0.5–1 μg. Always use an appropriate control antibody. An antibody targeting a protein known not to be present in the lysate is a good control for background signals.
4. Beads for different samples but the same antibody should always be incubated together so that the same conditions are applied to all samples.
5. The incubation time has to be individually tested for each antibody. Too long incubation can increase background signals.
6. The volume of the lysis buffer might have to be adjusted based on the amount of cells/tissue being used.
7. If working with small amounts of tissue or low numbers of cells, it is recommended to use fresh material. For bigger tissue pieces and higher cell numbers (>20 M) the starting material can be snap-frozen and kept on −80 °C until further use. However, it is recommended to test whether this is possible without RNA loss. If frozen material should be used for AGO-RIP, the cells/tissue should not be thawed before lysis, to avoid RNA loss. Instead add the lysis buffer and the steel bead quickly to the sample and immediately lyse.

8. Be careful not to touch the steel beads to avoid RNase contamination.
9. The amount of used lysis buffer depends on the amount of starting material and has to be adjusted. If there is a lot of foam after lysis, increase the buffer amount.
10. If there are tissue pieces left in the tube that are not homogenized properly, repeat this step. Put the homogenized lysate on ice until further use.
11. For some tissues it is necessary to repeat the centrifugation step to further clear the lysate.
12. The incubation time has to be optimized for the respective antibody and tissue.
13. Conduct DNAse digest as indicated in the kit protocol. Use isopropanol in the washing buffer if working with very low concentrations of RNA. In the elution step, add water and wait for 10 min before eluting, as this can increase the RNA concentration.
14. Do not touch the beads with the pipette. Place the pipette above the beads and gently rinse them off the tube wall.
15. If sequencing is to be conducted, it is recommended to take 2–3 μl of the RNA for cDNA synthesis, since the end material of the RIP is very little (14 μl) when using the miRNeasy micro kit.
16. As soon as cDNA is generated, it should be handled on a different bench than the RNA, with a separate set of pipettes.
17. Always include a water control sample.
18. The kit suggested in this protocol is a cDNA library preparation kit for total RNA sequencing. Alternatively a kit including poly-A enrichment can be used. However, the step of poly-A enrichment can lead to loss of RNA which can be problematic due to the commonly low RNA concentrations of RIP samples. Moreover, poly-A enrichment will exclude noncoding RNAs without poly-A tail.

References

1. Bartel DP (2009) MicroRNAs: target recognition and regulatory functions. Cell 136 (2):215–233. https://doi.org/10.1016/j.cell.2009.01.002. S0092-8674(09)00008-7 [pii]
2. Bartel DP (2004) MicroRNAs: genomics, biogenesis, mechanism, and function. Cell 116 (2):281–297
3. Krol J, Loedige I, Filipowicz W (2010) The widespread regulation of microRNA biogenesis, function and decay. Nat Rev Genet 11 (9):597–610. https://doi.org/10.1038/nrg2843
4. Kim VN (2005) MicroRNA biogenesis: coordinated cropping and dicing. Nat Rev Mol Cell Biol 6(5):376–385. https://doi.org/10.1038/nrm1644
5. Ebert MS, Sharp PA (2012) Roles for microRNAs in conferring robustness to biological processes. Cell 149(3):515–524. https://doi.org/10.1016/j.cell.2012.04.005
6. Guo H, Ingolia NT, Weissman JS, Bartel DP (2010) Mammalian microRNAs predominantly act to decrease target mRNA levels.

Nature 466(7308):835–840. https://doi.org/10.1038/nature09267
7. Hammell M, Long D, Zhang L, Lee A, Carmack CS, Han M, Ding Y, Ambros V (2008) mirWIP: microRNA target prediction based on microRNA-containing ribonucleoprotein-enriched transcripts. Nat Methods 5(9):813–819. https://doi.org/10.1038/nmeth.1247
8. Ritchie W, Flamant S, Rasko JE (2009) Predicting microRNA targets and functions: traps for the unwary. Nat Methods 6(6):397–398. https://doi.org/10.1038/nmeth0609-397
9. Boudreau RL, Jiang P, Gilmore BL, Spengler RM, Tirabassi R, Nelson JA, Ross CA, Xing Y, Davidson BL (2014) Transcriptome-wide discovery of microRNA binding sites in human brain. Neuron 81(2):294–305. https://doi.org/10.1016/j.neuron.2013.10.062
10. Chi SW, Zang JB, Mele A, Darnell RB (2009) Argonaute HITS-CLIP decodes microRNA-mRNA interaction maps. Nature 460(7254):479–486. https://doi.org/10.1038/nature08170
11. He M, Liu Y, Wang X, Zhang MQ, Hannon GJ, Huang ZJ (2012) Cell-type-based analysis of microRNA profiles in the mouse brain. Neuron 73(1):35–48. https://doi.org/10.1016/j.neuron.2011.11.010
12. Helwak A, Tollervey D (2014) Mapping the miRNA interactome by cross-linking ligation and sequencing of hybrids (CLASH). Nat Protoc 9(3):711–728. https://doi.org/10.1038/nprot.2014.043
13. Malmevik J, Petri R, Klussendorf T, Knauff P, Akerblom M, Johansson J, Soneji S, Jakobsson J (2015) Identification of the miRNA targetome in hippocampal neurons using RIP-seq. Sci Rep 5:12609. https://doi.org/10.1038/srep12609
14. Petri R, Pircs K, Jonsson ME, Akerblom M, Brattas PL, Klussendorf T, Jakobsson J (2017) let-7 regulates radial migration of new-born neurons through positive regulation of autophagy. EMBO J. 10.15252/embj.201695235
15. Hammond SM, Boettcher S, Caudy AA, Kobayashi R, Hannon GJ (2001) Argonaute2, a link between genetic and biochemical analyses of RNAi. Science 293(5532):1146–1150. https://doi.org/10.1126/science.1064023
16. Martin M (2011) Cutadapt removes adapter sequences from high-throughput sequencing reads. EMBnet J 17(1):10–12
17. Dobin A, Davis CA, Schlesinger F, Drenkow J, Zaleski C, Jha S, Batut P, Chaisson M, Gingeras TR (2013) STAR: ultrafast universal RNA-seq aligner. Bioinformatics 29(1):15–21. https://doi.org/10.1093/bioinformatics/bts635
18. Liao Y, Smyth GK, Shi W (2014) featureCounts: an efficient general purpose program for assigning sequence reads to genomic features. Bioinformatics 30(7):923–930. https://doi.org/10.1093/bioinformatics/btt656
19. Kozomara A, Griffiths-Jones S (2014) miRBase: annotating high confidence microRNAs using deep sequencing data. Nucleic Acids Res 42(Database issue):D68–D73. https://doi.org/10.1093/nar/gkt1181
20. Kersey PJ, Allen JE, Armean I, Boddu S, Bolt BJ, Carvalho-Silva D, Christensen M, Davis P, Falin LJ, Grabmueller C, Humphrey J, Kerhornou A, Khobova J, Aranganathan NK, Langridge N, Lowy E, McDowall MD, Maheswari U, Nuhn M, Ong CK, Overduin B, Paulini M, Pedro H, Perry E, Spudich G, Tapanari E, Walts B, Williams G, Tello-Ruiz M, Stein J, Wei S, Ware D, Bolser DM, Howe KL, Kulesha E, Lawson D, Maslen G, Staines DM (2016) Ensembl Genomes 2016: more genomes, more complexity. Nucleic Acids Res 44(D1):D574–D580. https://doi.org/10.1093/nar/gkv1209
21. Rosenbloom KR, Armstrong J, Barber GP, Casper J, Clawson H, Diekhans M, Dreszer TR, Fujita PA, Guruvadoo L, Haeussler M, Harte RA, Heitner S, Hickey G, Hinrichs AS, Hubley R, Karolchik D, Learned K, Lee BT, Li CH, Miga KH, Nguyen N, Paten B, Raney BJ, Smit AF, Speir ML, Zweig AS, Haussler D, Kuhn RM, Kent WJ (2015) The UCSC Genome Browser database: 2015 update. Nucleic Acids Res 43(Database issue):D670–D681. https://doi.org/10.1093/nar/gku1177
22. O'Leary NA, Wright MW, Brister JR, Ciufo S, Haddad D, McVeigh R, Rajput B, Robbertse B, Smith-White B, Ako-Adjei D, Astashyn A, Badretdin A, Bao Y, Blinkova O, Brover V, Chetvernin V, Choi J, Cox E, Ermolaeva O, Farrell CM, Goldfarb T, Gupta T, Haft D, Hatcher E, Hlavina W, Joardar VS, Kodali VK, Li W, Maglott D, Masterson P, McGarvey KM, Murphy MR, O'Neill K, Pujar S, Rangwala SH, Rausch D, Riddick LD, Schoch C, Shkeda A, Storz SS, Sun H, Thibaud-Nissen F, Tolstoy I, Tully RE, Vatsan AR, Wallin C, Webb D, Wu W, Landrum MJ, Kimchi A, Tatusova T, DiCuccio M, Kitts P, Murphy TD, Pruitt KD (2016) Reference sequence (RefSeq) database at NCBI: current status, taxonomic expansion, and functional annotation. Nucleic Acids Res 44(D1):D733–D745. https://doi.org/10.1093/nar/gkv1189

Chapter 10

Integrated Analysis of miRNA and mRNA Expression Profiles to Identify miRNA Targets

Zhiming Li and Chi-Meng Tzeng

Abstract

miRNAs are involved in various biological processes and different diseases through interacting with their target mRNAs. Therefore, deciphering miRNA targets is important for diagnostics and therapeutics. However, it is still complex and expensive to do large scale screening for miRNA alterations and identify their downstream miRNA targets. Recently, integrating analysis of miRNA expression profiles and mRNA expression levels has been used to successfully identify the most prominent interactions. Here, we present a protocol that combines both expression data and computational prediction to refine the microarray result for identifying the miRNA targets. Many details are also provided to help researchers choose more appropriate tools and methods for a given type of application. Available web-based resources and experimental conditions for functional enrichment and validation of miRNA targets are also summarized.

Key words Integrated analysis, miRNA target, mRNA, Enrichment, Microarray

1 Introduction

A number of miRNA therapeutics are currently in clinical trials. A phase I clinical trial in which a Locked Nucleic Acid (LNA)-based anti-miRNA targeting miR-122 was used to treat hepatitis C has recently been completed with encouraging results [1]. Clinical trials of selected miRNAs, including let-7, miR-29, miR-21, and miR-34, suggest these miRNAs are suitable potential clinical targets in cancer [2]. This demonstrates the possibility of miRNAs as therapeutic targets and miRNA inhibitors as well as mimics as a new class of drugs. Further improvement of the biotechnology into a successful therapy involves two key aspects: developing efficient in vivo delivery systems and identifying new miRNA targets.

Putative identification using seed sequence complementarity and free energy predictions of miRNA–mRNA duplexes are available in databases such as TargetScan. However, the false positive rate for such matches is unacceptably high. We have developed and validated a simple method to identify functional mRNA targets of

Shireen R. Lamandé (ed.), *mRNA Decay: Methods and Protocols*, Methods in Molecular Biology, vol. 1720,
https://doi.org/10.1007/978-1-4939-7540-2_10,

miRNA from microarray data [3]. The method looks for the subset of anticorrelated miRNA–mRNA pairs from a larger set of miRNA and mRNA differently expressed in cases and controls. Our method can not only identify miRNA–mRNA pairs which discriminate cases from controls, but also predict the dysregulated regulation mechanisms between them that may initiate and/or drive the disease process using GO and KEGG. Its principal merit is the ability to reduce the large number of relatively speculative matches from seed sequences alone to a smaller set of functional, tissue specific targets.

Recently, when microarrays were used to measure the expression levels of miRNAs and mRNAs simultaneously, integrated analysis was extensively performed in the search for possible candidate miRNAs and their mRNA targets in various diseases and fields, including azoospermia [3], aggressive papillary thyroid carcinoma [4], atrial fibrosis [5], colorectal cancer [6], and rice seedlings [7]. Clearly the miRNA and its target mRNA must be coexpressed in order for the miRNA to repress the expression of its biological target. A straight forward method to analyze the relationship between miRNAs and mRNAs is the Pearson correlation. Alterations in miRNA regulatory pathways can cause different diseases. Enrichment analysis of miRNA targets is crucial and helpful to elucidate underlying changes of miRNA functions. A biological basis of enrichment analysis is that genes interact to form a module or interaction network. In order to help researchers choose more appropriate tools for a given type of analysis, we describe in the following sections the capabilities of these tools and the experimental methods employed in detail.

2 Materials

2.1 Computational Tools

1. Hardware (64-bit computer; 4 GB of RAM (16 GB preferred)).
2. Windows operating system, with access to the Internet.
3. Depending on the experimental goals and selection of analytic tools the R programming language (available from https://www.r-project.org/) or Java Runtime Environment (available from http://www.java.com) locally.

2.2 Microarray

1. Agilent Human miRNA Microarray 8 × 60K Release 16.0 (Agilent Technologies).
2. Agilent SurePrint G3 Human gene expression 8 × 60K microarray (Agilent Technologies).
3. miRNeasy Micro Kit (Qiagen).
4. TRIzol reagent (Invitrogen).

5. Microarray scanner.
6. GeneSpring GX software version 12.0 (Agilent Technologies).

2.3 Quantitative RT PCR

1. 10 × miScript Primer Assays (Qiagen).
2. RNU6B snRNA primer assay (Qiagen).
3. Quanti-Tect Reverse Transcription Kit (Qiagen).
4. TransStar Top Green qPCR Super Mix (Transgene).

2.4 Molecular Cloning

1. TransStart FastPfu DNA polymerase (Transgene).
2. pmirGlo Dual-Luciferase miRNA Target Expression Vector (Promega).
3. Restriction enzymes (depending on the insert).
4. T4 DNA ligase and T4 ligase buffer.
5. Distilled water (dH_2O).

2.5 Cell Culture and Transfection

1. Dulbecco's modified Eagle's medium (DMEM) supplemented with 10% fetal bovine serum.
2. Phosphate-buffered saline (PBS).
3. Solution of 0.25% trypsin and 1 mM EDTA.
4. Lipofectamine 2000 (Invitrogen).

2.6 Luciferase Assay

1. Transfected and incubated HEK 293 cells.
2. Dual-Luciferase Reporter Assay Kit (Promega).

2.7 Immunoblotting

1. Radioimmunoprecipitation assay (RIPA) lysis buffer.
2. 1× Complete EDTA-free protease inhibitors (Roche).
3. Polyvinylidene difluoride membrane.
4. ECL detection system.

3 Methods

3.1 Microarray Data Analysis

1. Follow the manufacturer's instructions for RNA hybridization and detecting fluorescent signal intensities with a microarray scanner. While dependent on the data source, our findings suggest that consistent results are typically obtained using three biological replicates for each group.
2. Normalize gene expression and miRNA microarray raw data using the GeneSpring GX software version 12.0. Transform signal values to log base 2, and then apply quantile and percentile shift to obtain an equal distribution of probe signal intensities.

3. Compare treatment and control groups using the *t*-test (*p*-values) and the Benjamini–Hochberg False Discovery Rate (FDR) correction (adjusted *p*-values). Genes and miRNAs are taken to be differentially expressed if the adjusted *p*-values < 0.05 and fold change (FC) is greater than 2 (upregulation) or less than $1/2$ (downregulation).

3.2 GO and KEGG Pathway Analysis

1. Use Gene Ontology (GO) and KEGG pathway analysis for all of the differentially expressed genes with Cytoscape V2.7 (http://cytoscape.org/) and the ClueGo V1.3 plug-in. ClueGO determines the distribution of the target gene list across the GO terms and pathways.
2. Calculate *p*-values using a right-side hypergeometric test, and Benjamini–Hochberg adjustment to correct for multiple tests. An adjusted *p*-value < 0.05 indicates a statistically significant deviation from the expected distribution, and thus the corresponding GO terms and pathways are enriched in target genes.

3.3 miRNA–mRNA Enrichment Analysis and Negative Correlation

1. Predict putative target genes of differentially expressed miRNAs with the following algorithms: MiRanda, Pictar, TargetScans, Microrna.org, PITA, MicroT, RNA22. All predicted results (union set) are subject to intersect with miRNA–mRNA pairs refined from microarray data in the **step 2** below.
2. Combine miRNA and mRNA microarray data to obtain reliable miRNA–mRNA pairs with significant Pearson correlation and negative correlation in the study.
   (a) Apply Pearson correlation analysis to the top 75% differentially expressed miRNAs and mRNAs in the microarray analysis, then calculate *p*-values using a right-side hypergeometric test, and Benjamini–Hochberg adjustment to correct for multiple tests for above miRNAs and mRNAs. An adjusted *p*-value < 0.05 indicates a statistically significant deviation from the expected distribution.
   (b) Compute Pearson correlation coefficients between the particular miRNA and its target mRNAs in the **step (a)** using R (http://www.R-project.org) to determine whether the expression levels of each miRNA and its mRNA targets are negatively correlated (correlation < 0 and Fisher *P* corrected by FDR < 0.05).
   (c) For each miRNA target found that meets the above two criteria, a variable theta (θ) indicates the strength of the relationship. $\theta = \text{FC1}/\text{FC2}$, where FC1 is the fold change of significantly expressed genes and FC2 is the fold change of significantly expressed miRNAs. Take the reciprocal of FC, if FC < 0. The greater the absolute value of θ, the stronger the relationship between mRNA and miRNAs.

3.4 Functional Network Analysis

1. Collect experimentally validated miRNA targets from TarBase database.
2. Download the human protein–protein interaction networks (PPIN) from the Human Protein Reference Database.
3. Integrate TarBase, PPIN, and predicted data (*see* in Subheading 3.3) using the igraph package of the statistical language R to make functional profiling. Network analysis and visualization can be applied by Cytoscape tool.
4. Use the Markov cluster algorithm to identify highly connected modules within the global networks.

3.5 Hierarchical Clustering

1. Perform hierarchical clustering in Cluster version 3.0 and Java TreeView version 1.1.6 to identify and visualize patterns within the dataset. With the use of clustering algorithms the samples and probes are grouped based on similarities in the expression profile.
2. Filter the probe set based on standard deviation to exclude the probes of least variance. Choose the parameters average linkage and median centering.
3. Use both unsupervised and supervised clustering. In the unsupervised method all genes are included, while supervised clustering involves input of just significant genes.
4. Commonly choose those highly differentially expressed genes and the genes that may be involved in biological progress according to the GO and KEGG analysis (*see* Subheading 3.2). Novel genes deserve more care and concern.

3.6 Quantitative Real-Time RT-PCR for miRNAs and mRNAs

1. Perform relative quantitative real time PCR to confirm the array results using the miScript PCR System along with the 10 × miScript Primer Assays for interesting miRNAs, along with the RNU6B snRNA primer assay as an endogenous reference for normalization, according to the manufacturer's recommendations.
2. Reverse transcribe 2 μg of total RNA from each sample using a reverse transcription kit according to the manufacturer's protocol. Amplify the reverse transcription products using the TransStar Top Green qPCR Super Mix according to the manufacturer's protocol. Include a suitable internal control such as GAPDH (*see* **Note 1**).
3. The results, analyzed by the comparative Ct method or 2-ΔΔCt method for miRNAs and mRNAs relative expression (corrected to a miRNA and mRNA reference, such as RNU6 and GAPDH, respectively), can be reported as the relative fold change versus the control sample. The log2 normalization of qPCR results was used for comparison with microarray data.

4. Currently there is no clear consensus as to what criteria should be followed to determine miRNA targets and to confirm their biological efficacy. Based on our previous experience with miRNAs [3, 8], we propose two additional criteria should be together met before further confirmation. First, miRNAs and target mRNAs should have both significantly differentiated expression (*see* Subheading 3.1). Second, mRNAs and target mRNAs should be listed in the intersection between bioinformatic prediction in silico analysis and statistic correlation in microarray analysis (*see* Subheading 3.3).

3.7 Plasmid Construction and Transfection

1. To test the effect of miRNA-mediated, posttranscriptional regulation on target genes engineer a firefly luciferase gene construct containing the predicted miRNA targeting sequence from the target gene (often located in the 3′-UTR). Common bioinformatics algorithms listed in the **step 1** of Subheading 3.3 are applied to search for complementary sequences in the 3′UTRs of potential target mRNAs (*see* **Note 2**).
2. Obtain the 3′-UTR of your target gene (including the seed sequence of the miRNA of interest) by PCR amplification using Pfu DNA polymerase. Note, the 3′-UTRs may contain multiple targeting sequences and other regulatory elements. Commonly specific PCR using custom forward and reverse primers can be used to obtain miRNA targeting sequences. The pmirGLO Dual-Luciferase miRNA Target Expression Vector (Promega, USA) is used to confirm the function of the putative miRNA binding sites in the 3′-UTR of target genes.
3. Cultivate the required number of HEK 293 cells (cells should be 70% confluent on the day of the experiment) for a 24-well plate, 1×10^5 cells per well are needed. Wash HEK 293 cells with serum-free medium once and then incubate with serum-free medium for 4–6 h. The recombinant pmirGLO plasmid and a miRNA of interest are then transiently transfected into a host cell, preferably one that does not endogenously express this miRNA.
4. 50 nM miRNA mimic (or 50 nM scrambled miRNA) and 0.5 μg pmirGLO, and 2 μL Lipofectamine 2000 are separately mixed with 25 μL Opti-MEM I Reduced Serum Medium for 15 min (*see* **Notes 3** and **4**). Then, the two mixtures are combined and incubated for 20 min at room temperature. The Lipofectamine 2000-miRNA mixture is added to the cells and incubated at 37 °C for 12 h. Subsequently, 0.5 mL fresh medium containing 10% fetal bovine serum is added to the flasks and the cells maintained in the culture until the experiment.

3.8 Dual Luciferase Reporter Gene Assay

1. Remove growth media from cultured cells. Rinse cultured cells in 1 × PBS. Then remove all rinse solution. Dispense the recommended 100 μL 1 × Passive Lysis Buffer (PLB) into each well of 24-well plate.
2. Gently rock/shake the culture vessel for 15 min at room temperature. Transfer lysate to the 96-well plate. Plate with about 20 μL of PLB Lysate/well. Luciferase and Renilla activity was measured using the Dual-Luciferase Reporter Assay Kit according to the manufacturer's instructions (*see* **Note 5**).

3.9 Immunoblotting

1. Cells, which endogenously express target protein, are transfected with the indicated miRNA mimics and inhibitors.
2. Forty-eight hours later, lyse cells using RIPA buffer containing 1 × Protease Inhibitor Cocktail.
3. Run 40 μg of total protein on an SDS/PAGE gel and transfer to PVDF membrane using the Bio-Rad V3 Western work flow system according to the manufacturer's recommendation.
4. Probe with diluted primary antibody to the target protein overnight at 4 °C. The secondary antibody should be a horseradish peroxidase (HRP)-conjugated IgG, whereas HRP-conjugated anti-β-tubulin antibody was used as the loading control (*see* **Note 6**). Detect the signal with an ECL detection system according to the manufacturer's instructions.

4 Notes

1. TaqMan MicroRNA Assay also could be applied to verify microarray result, although the example data that we described were obtained using SYBR-based miRNA assay in this protocol.
2. While most studies have focused on miRNA targeting sites located in the 3′-UTR, a number of miRNA targeting sites have also been found in the coding region. Although our method presented here is for testing the miRNA targeting site located in the 3′-UTR of the targeted mRNA, it can be adapted to test miRNA targeting sites in the coding regions.
3. When using RNA and miRNA mimic/inhibitor pay attention to work with RNase-free equipment and RNase-free solutions. For long-term storage of your RNA extracts it is best to keep them precipitated at −80 °C.
4. To analyze the regulation of gene expression by miRNAs, four different transfection experiments have to be performed for each target region: (1) miRNA and the pmirGLO harboring the putative binding site, (2) miRNA negative control and the pmirGLO harboring the putative binding site, (3) miRNA and

the pmirGLO harboring the mutated binding site, and (4) miRNA negative control and the pmirGLO harboring the mutated binding site.

5. To confirm whether the miRNA of choice really affects the expression of its targeted mRNA, harvest the cell lines that overexpress the miRNA and subject the cells to immunoblot blot analysis of the protein level of target gene.
6. The limitation of the method as presented is that it cannot find targets that are repressed by translation inhibition. For such targets, the alterations in miRNA levels would leave mRNA levels unaffected, but decrease protein levels.

Acknowledgment

This work was supported by the Natural Science Foundation of Fujian Province (No. 2016J01171) of China.

References

1. Biswas S, Roy S, Banerjee J, Hussain SR, Khanna S, Meenakshisundaram G, Kuppusamy P, Friedman A, Sen CK (2010) Hypoxia inducible microRNA 210 attenuates keratinocyte proliferation and impairs closure in a murine model of ischemic wounds. Proc Natl Acad Sci U S A 107(15):6976–6981. https://doi.org/10.1073/pnas.1001653107
2. Bader AG (2012) miR-34 – a microRNA replacement therapy is headed to the clinic. Front Genet 3:120. https://doi.org/10.3389/fgene.2012.00120
3. Zhuang X, Li Z, Lin H, Gu L, Lin Q, Lu Z, Tzeng CM (2015) Integrated miRNA and mRNA expression profiling to identify mRNA targets of dysregulated miRNAs in non-obstructive azoospermia. Sci Rep 5:7922. https://doi.org/10.1038/srep07922
4. Yang Z, Yuan Z, Fan Y, Deng X, Zheng Q (2013) Integrated analyses of microRNA and mRNA expression profiles in aggressive papillary thyroid carcinoma. Mol Med Rep 8 (5):1353–1358. https://doi.org/10.3892/mmr.2013.1699
5. Wang J, Wang Y, Han J, Li Y, Xie C, Xie L, Shi J, Zhang J, Yang B, Chen D, Meng X (2015) Integrated analysis of microRNA and mRNA expression profiles in the left atrium of patients with nonvalvular paroxysmal atrial fibrillation: role of miR-146b-5p in atrial fibrosis. Heart Rhythm 12(5):1018–1026. https://doi.org/10.1016/j.hrthm.2015.01.026
6. Vishnubalaji R, Hamam R, Abdulla MH, Mohammed MA, Kassem M, Al-Obeed O, Aldahmash A, Alajez NM (2015) Genome-wide mRNA and miRNA expression profiling reveal multiple regulatory networks in colorectal cancer. Cell Death Dis 6:e1614. https://doi.org/10.1038/cddis.2014.556
7. Tang M, Mao D, Xu L, Li D, Song S, Chen C (2014) Integrated analysis of miRNA and mRNA expression profiles in response to Cd exposure in rice seedlings. BMC Genomics 15:835. https://doi.org/10.1186/1471-2164-15-835
8. Li Z, Zheng Z, Ruan J, Li Z, Zhuang X, Tzeng CM (2016) Integrated analysis miRNA and mRNA profiling in patients with severe oligozoospermia reveals miR-34c-3p downregulates PLCXD3 expression. Oncotarget 7 (33):52781–52796. 10.18632/oncotarget.10947

Chapter 11

Identifying RISC Components Using Ago2 Immunoprecipitation and Mass Spectrometry

Tingfang Yi

Abstract

Complex immunoprecipitation (Co-IP) is a powerful technique for precipitating an intact protein complex out of solution and cell lysates using an antibody that specifically binds to a particular protein in a large complex of proteins. Mass spectrometry (MS) is used to identify, sequence, and quantify proteins. RNA-induced silencing complexes (RISCs), Ago2 centered protein assemblies, are essential for miRNA mediated RNA decay and gene expression regulation; however, the complete list of RISCs is unknown. Here we describe methods used to combine IP and MS to identify new components of RISCs.

Key words RISC, Ago2, Component, Complex immunoprecipitation, Mass spectrometry

1 Introduction

The tiny miRNAs regulate genome expression and mediate development, and aberrant miRNA processes induce various diseases including cancer. RNA-induced silencing complexes (RISCs) play essential roles in miRNA-mediated mRNA decay, degradation, and translation regulation. Argonaute-2 (Ago2) is a central scaffold of RISC. Multiple proteins are documented to be RISC components, including Ago2, Dicer, TRBP, ADAR1 [1], PARN [2], RNA helicase A [3], Hsp90 [4], TSN [5], and eIF1A [6]. However, the complete list of RISC components is unknown.

Immunoprecipitation (IP) is a technique for precipitating a specific protein antigen out of solution (or cell lysates) using an antibody bound to solid beads that specifically binds to the particular protein [7]. Complex immunoprecipitation (Co-IP) works by selecting an antibody targeting a known component of a large complex of proteins [8]. After the specific intact protein complex is precipitated and isolated, the complex components can be eluted and collected for protein identification. Gel electrophoresis can separate proteins based on their size and charge. Highly sensitive silver staining can detect 0.25 ng of protein in the SDS-gel. In-gel

Shireen R. Lamandé (ed.), *mRNA Decay: Methods and Protocols*, Methods in Molecular Biology, vol. 1720,
https://doi.org/10.1007/978-1-4939-7540-2_11,

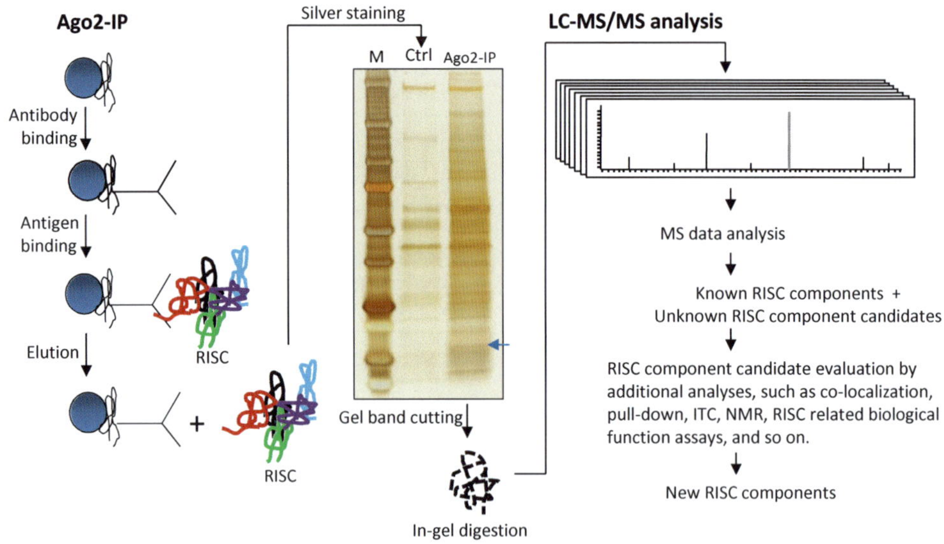

Fig. 1 Schematic diagram of Ago2-IP and LC-MS/MS methods for identifying new RISC components. The antibody bound solid beads targeting Ago2 immunoprecipitate intact RISC complexes. Gel electrophoresis separates RISC complex components and silver staining displays distinct component bands for LC-MS/MS analysis. The MS data analysis and post-MS assays identify new RISC components. The *blue arrow* shows the protein band location of eIF1A, a component of RISC, in a silver staining gel

digestion and peptide extraction can remove low molecular weight impurities which are often detrimental for mass spectrometric sequencing [9]. Mass spectrometry, which ionizes polypeptides and separates ions based on mass to charge ratio, is a powerful technique for identifying, sequencing, and quantifying polypeptides [10–12].

Thus, combining Ago2-IP, MS, and post-MS analyses (such as NMR [13–15]), we have identified a new RISC component, eIF1A, which directly interacts with the MID-domain of Ago2 and functions in DICER-independent miRNA biogenesis and miRNA-mediated RNA interference (RNAi) (Fig. 1). The combination of Ago2-IP and MS methods helps to identify new components in RISCs, the key protein complexes in miRNA mediated gene expression regulation.

2 Materials

2.1 Immunoprecipitation

1. 1× PBS (Phosphate-buffered saline): 137 mM NaCl, 2.7 mM KCl, 4.3 mM Na_2HPO_4, 1.47 mM KH_2PO_4, pH 7.4.
2. Native cell lysis buffer: 20 mM Tris–HCl, pH 7.4, 137 mM NaCl, 10% glycerol, 1% NP-40, with protease inhibitors. Water should be Milli-Q water (*see* **Note 1**).
3. Ribonuclease: Ribonuclease A and RNAase (*see* **Note 2**).

4. Antibodies: targeted protein specific antibodies.
5. IP kit Dynabeads Protein G.
6. Magnet.

2.2 Silver Staining

1. Fixative: 40% (v/v) methanol, 10% (v/v) acetic acid. For 200 ml fixative: 80 ml methanol (99.8%), 20 ml acetic acid (100%), 100 ml H_2O.
2. Wash solution: 200 ml 35% (v/v) ethanol.
3. Sensitizing solution: 200 ml 0.02% $Na_2S_2O_3$.
4. Silver staining reagent: 0.2% $AgNO_3$, 0.076% formalin (35% Formaldehyde). For 200 ml (0.4 g $AgNO_3$, 152 μl formalin (35% Formaldehyde), adjust to final 200 ml with H_2O).
5. Developer: 6% Na_2CO_3, 0.05% formalin (35% Formaldehyde), 0.0004% $Na_2S_2O_3$. For 400 ml developer: 24 g Na_2CO_3, 200 μl formalin (35% Formaldehyde), 8 ml 0.02% $Na_2S_2O_3$, and 392 ml H_2O.
6. Stop solution: 5% (v/v) acetic acid. 200 ml (10 ml acetic acid, 190 ml H_2O).

2.3 In-Gel Digestion for MS Analysis

1. 100 mM Ammonium bicarbonate.
2. Dithiothreitol (DTT).
3. Iodoacetamide.
4. Trypsin (sequencing grade, modified), 13 ng/μl trypsin in 10 mM ammonium bicarbonate containing 10% (v/v) acetonitrile. Use $H_2{}^{18}O$ instead of normal $H_2{}^{16}O$ for peptide quantification or de novo sequencing.
5. Water (LiChrosolv grade).
6. Acetonitrile (HPLC gradient grade).
7. Formic acid (reagent grade).
8. Trifluoroacetic acid (Uvasol grade).

3 Methods

Carry out all procedures at room temperature unless otherwise specified.

3.1 Ago2-IP

1. Wash the cultured adherent cells (if suspended cells, spin down the cells) with 1× PBS twice to remove the chemicals in the culture media (*see* **Note 3**).
2. Add cold native cell lysis buffer (on ice) onto the cells (1 ml for a 10 cm dish) and immerse all cells. Scrape the cell lysates with a clean rubber scraper and transfer all the cell lysate into eppendorf tubes.

3. Incubate the tubes with rotation for 5 min at 4 °C to sufficiently lysate the membranes. Sonicate the cell lysates with EpiShear Probe Sonicator for 3 × 5 s (on ice) to shear DNAs and RNAs while do not do harm to RISC assembly (*see* **Note 4**). The sheared DNAs/RNAs will reduce nonspecific binding proteins. Rotate the tubes for 3 min at 4 °C. Spin the cell lysates with high speed (13,523 rcf (*g*)) for 5–10 min. Transfer the upper liquid into a new tube.
4. Add Ribonuclease A and RNAase (*see* **Notes 5** and **6**). Keep the samples at room temperature for 30 min to digest DNAs and RNAs. The digested DNAs/RNAs will reduce nonspecific binding proteins. Put the sample tubes on ice.
5. Dynabead preparation (*see* **Note 7**). Resuspend Dynabeads in the vial and transfer 50 μl Dynabeads to a tube. Place the tube on the magnet to separate the beads from the solution, and remove the supernatant. Remove the tube from the magnet.
6. Antibody binding (*see* **Note 8**). Add antibody solution (2–10 μg antibodies in 200 μl 1× PBS with 0.01% Tween 20®) to the tube with Dynabeads. Incubate with rotation for 10 min at room temperature (*see* **Note 9**). Place the tube on the magnet and remove the supernatant. Add 200 μl 1× PBS with 0.01% Tween-20®), and suspend the Dynabeads-Ab (antibody) complex with gentle pipetting.
7. Immunoprecipitation. Place the tube containing Dynabeads-Ab complex on magnet and remove the supernatant. Remove the tube from the magnet. Add the cell lysate supernatants containing the antigen (Ag) (100–1000 μl) and gently pipette to resuspend the Dynabeads-Ab complex. Incubate with rotation for 20 min or 1 h at 4 °C to allow antigen bind the Dynabeads-Ab complex (*see* **Note 10**).
8. Elution. Place the tube on magnet and transfer the supernatant to a new tube for further analysis if desired. Wash the Dynabeads-Ab complex three times using 200 μl washing buffer for each wash (move the tube from magnet when add washing buffer and gently suspend the Dynabeads-Ab complex) (*see* **Note 11**). Add 20–30 μl RIPA cell lysis buffer (with 1% SDS) and suspend the Dynabeads-Ab complex (avoid foaming). Incubate with rotation for 5 min or directly boil the sample-beads solution for 5 min. Place tube on magnet, transfer the supernatant to a new tube (Ago2 precipitated RISC components are in the supernatant).

3.2 Silver Staining

1. SDS gel electrophoresis (*see* **Notes 12** and **13**). Gel electrophoresis removes low molecular weight impurities, including detergents, buffer components, which are often detrimental for mass spectrometric sequencing. Prerun the Mini-PROTEAN

TGX Precast Gels with 1× SDS running buffer without sample for 20 min, which will remove the nonpolymerized chemicals in the gel. Add the Ago2-IP samples to the Mini-PROTEAN TGX Precast Gels and run gel electrophoresis in the 1× SDS running buffer.

2. After SDS gel electrophoresis, transfer the gel into a clean glass container. Wash the gel with water three times to remove free chemicals.
3. Fix gel: 40% methanol, 10% acetic acid, 1 h.
4. Wash gel: deionized water (MilliQ water) for 20 min.
5. Wash gel: deionized water for 20 min.
6. Wash gel: deionized water for 20 min. The several washes will remove all acetic acid, reduce background staining and increase sensitivity.
7. Sensitize gel: 0.02% $Na_2S_2O_3$ for 1–2 min.
8. Wash gel: deionized water for 5 min.
9. Wash gel: deionized water for 5 min.
10. Wash gel: deionized water for 5 min.
11. Stain gel: Stain the gel in 0.1% $AgNO_3$, 0.02% formaldehyde solution (0.2 g $AgNO_3$, 200 ml H_2O), 0.02% formaldehyde (add 40 μl 35% formaldehyde just before use) for 20 min.
12. Wash gel: deionized water for 1 min.
13. Wash gel: deionized water for 1 min.
14. Develop gel: 3% Na_2CO_3, 0.05% formalin (7.5 g Na_2CO_3 in 250 ml H_2O), 0.05% formaldehyde (add 125 μl 35% formaldehyde just before use). Change developer solution immediately when it turns yellow. Terminate when the staining is sufficient (*see* **Note 14**).
15. Stop staining: 5% acetic acid for 5 min.
16. Leave the gel in 1% acetic acid at 4 °C for storage. Prior to MS analysis the gel is washed in water for 3 × 10 min to ensure complete removal of acetic acid.

3.3 In-Gel Digestion for MS Analysis

1. Excise protein bands (*see* **Notes 15** and **16**). Rinse the entire gel with deionized water for 1–2 h, put a plastic tray with the gel onto a light box and excise bands of interest with a clean scalpel.
2. Transfer the bands into tubes. Spin the gel pieces down to the tube bottom.
3. In-gel reduction, alkylation, and destaining of proteins. Add 500 μl of neat acetonitrile and incubate tubes for 10 min until gel pieces shrink, become opaque and stick together. Spin the gel pieces down, remove all liquid.

4. Add 50 μl DTT solution to cover the gel pieces. Incubate in air thermostat at 56 °C for 30 min. Cool down the tubes to room temperature, add 500 μl acetonitrile and incubate for 10 min, remove all the liquid (*see* **Note 17**).
5. Add 50 μl of iodoacetamide solution and incubate in the dark for 20 min.
6. Shrink gel pieces with acetonitrile for 10 min, and remove all liquid.
7. Add 50 μl trypsin solution to cover the dry gel pieces. Place the tube in ice bucket or a fridge for 30 min.
8. Check if all the trypsin solution was absorbed and add more trypsin solution, if necessary. All gel pieces should be covered by trypsin solution.
9. Saturate the gel pieces for 90 min. Add 20 μl 100 mM ammonium bicarbonate (*see* **Note 18**) to cover the gel pieces and keep them wet during enzymatic cleavage.
10. Digestion. Place the tubes containing the gel pieces and enzyme solution in an air circulation thermostat and incubate at 37 °C overnight. To prevent condensation of water in the tube which leads to premature dehydration of the gel pieces, it is important to avoid a temperature gradient between the tube bottom and lid.
11. Chill tube to room temperature, spin down the gel pieces in a microcentrifuge.
12. Withdraw 1–1.5 μl aliquot of the supernatant from the digest for LC-MS/MS analysis. The left supernatant containing the digests can be stored at −20 °C for a few months (the gel pieces are in the tube bottom).
13. Peptide extraction. Add 100 μl of extraction buffer (1:2 (v/v) 5% formic acid/acetonitrile) to the tube containing digests and incubate for 15 min at 37 °C in a shaker.
14. Collect the supernatant into a PCR tube, dry down in a vacuum centrifuge. Keep the gel pieces in the tube and do not discard extracted gel pieces. The dried extracts can be safely stored at −20 °C for a few months for LC-MS/MS analysis.

3.4 MS and Post-MS Analysis

1. The spectral data can be searched with SEQUEST against a database containing the human International Protein Index protein sequence database: (http://www.ensembl.org/index.html).
2. The data of peptide sequences and intensity are analyzed (*see* **Note 19**). Usually a gel band containing a single protein lane shows dozens of or even hundreds of proteins in the MS data set. Normally the protein with the highest peptide intensity and

most polypeptide fragments is the dominant protein in the band/lane, which is pulled down in the Ago2 IP.

3. Search the biological functions of the interested proteins in PubMed, analyze the known and predicted protein–protein interaction with http://string-db.org/. Analyze the structure of interesting proteins in PDB (http://www.rcsb.org/pdb/home/home.do), if they are available. Shorten the RISC component candidate list.

4. It is important to exclude the nonspecific binding proteins and indirect interacting proteins in identifying new RISC components. This needs additional tests to determine the direct interaction between RISC components and the biological functions of the candidates in RISC procedures (*see* **Note 20**). There are several powerful techniques and methods for identifying protein–protein interactions including nuclear magnetic resonance (NMR) (Fig. 2), Isothermal titration calorimetry (ITC), colocalization assay with fluorescence microscope spectrometry in cells, GST-pull down assay, and so on. The verified protein–protein interaction between a candidate and the known RISC components and the specific biological functions of a candidate in RISC procedures can support a candidate as a new RISC component.

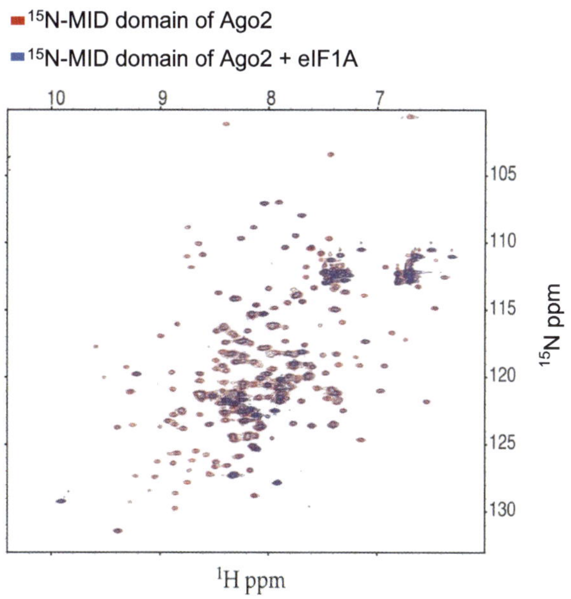

Fig. 2 An NMR showing that human eIF1A binds to the MID-domain of Ago2 in a RNA-binding-independent manner. ^{1}H-^{15}N TROSY-HSQC spectrum of ^{15}N-Sumo-MID (in *red*) titrated with eIF1A (in *blue*). The titration of eIF1A (molar ratio of 1:1) results in broadening ^{1}H-^{15}N TROSY-HSQC crosspeaks coming from the ^{15}N-MID domain, but not from the ^{15}N-Sumo (SMT3) tag

4 Notes

1. Using the native lysis buffer without SDS during the lysis procedure helps to protect the complex intact. Gentle pipetting during all the Co-IP processes avoids physical harm to the RISC complex assembly.
2. Proteinase inhibitors help to prevent protein degradation. Phosphatase inhibitors inactivate phosphatases avoiding dephosphorylation caused protein–protein interaction disruption in RISC assembly.
3. Wear gloves during all procedures.
4. Brief microsonication shears chromatin, DNA and RNA, which helps to decrease soluble protein loss, increase IP products, and reduce nonspecific binding proteins during Co-IP. Shearing DNA/RNA also increases the quality of silver staining.
5. DNAases degrade DNAs, which further help to reduce nonspecific binding proteins and increase Co-IP quality and products.
6. RNAases degrade RNAs, which help to avoid nonspecific binding linked by mRNAs, noncoding RNAs, and pre-miRNAs and miRNAs. As mature RISCs incorporated with miRNA and mRNA, and many factors beyond RISCs interact with mRNAs in the cytoplasm and nucleus. It is important to decrease long mRNAs to the minimum level to obtain non-RNA-induced protein interaction in isolating and identifying unknown RISC components.
7. Advantages of magnet beads: (1) high efficiency, the Dyna-deads in the magnet field are collected on the side of the tube, the washing supernatant are completely removed, which avoids intense pipetting with agarose beads, and decreases the nonspecific binding proteins; the complete procedure is approximately 30 min, decreasing the physical disruption effects on the complex; (2) easy to handle; (3) increase Co-IP quality.
8. Pay attention to antibody specificity, concentration, and quality during the Ago2-IP. Antibodies with high specificity and good quality help to reduce nonspecific binding proteins. Antibody quantity can be adjusted according to the protein quantity of the samples.
9. During the antibody binding and antigen binding procedures, incubation time can be increased. On the other hand, extensive antigen binding times will increase the nonspecific binding products, but cannot significantly increase the RISC component products.

10. Incubation temperature during antigen binding. Place the tubes in cold room at 4 °C with rotation, which helps to reduce protein degradation by proteases in the cell lysates that have not been inactivated.
11. Decreasing the nonspecific binding proteins helps to reduce post-MS analysis work and increase the RISC component identification efficiency. It is important to completely remove all the supernatant during the three washes with sufficient washing buffer (at least 20 times the volume of beads each time).
12. Precast gels are safe, ready for use, time saving and convenient. Unpolymerized acrylamide is a suspected carcinogen and neurotoxin, and may cause reproductive toxic effects. If you make the SDS gel yourself, unpolymerized powder and solutions of acrylamide should be handled with great care. A mask, lab coat, and a pair of nitrile gloves must be worn when weighing acrylamide in a fume hood. Avoid exposing acrylamide to any person during the weighing, transferring and dissolving. Avoid contact of unpolymerized acrylamide solutions with skin because skin contact causes toxic and systemic effects. Do not inhale acrylamide powder.
13. Silver staining has high sensitivity and can detect 0.25 ng of protein, while classical Coomassie blue staining can detect only detect 100 ng of protein. The higher sensitivity of silver staining helps to detect more bands and increase protein specificity in the gel bands.
14. When the gel turns light yellow, immediately add stop buffer. The gel color will become darker after adding stop buffer because the development termination takes about 30 s. If the gel turns into clear yellow, even when stop buffer is added quickly, the gel will become dark yellow or dark, which increases the difficulty of band distinction and gel excision.
15. Take special care to avoid keratin contamination of the samples when excising the gel bands.
16. Take an image of the silver stained gel and print the image before gel excision. Label each tube with the gel band location and order to prevent gel band confusion and mistakes.
17. Pay special care to avoid contamination during the in-gel digestion. Make the trypsin solution shortly before use and discard unused volume.
18. Make ammonium bicarbonate solution daily and discard after use. Make 10 mM DTT in 100 mM ammonium bicarbonate shortly before use.
19. Peptide intensity and fragment numbers reflect the protein quantity in the single excised gel band to some degree. However, sometimes several proteins show very similar high peptide

intensity and multiple fragment numbers in a single band. The known biological functions, the full length polypeptide sequence, the protein structure (if it is available), published data showing the protein is pulled down by other known RISC components, structure–function relationships, predicted protein–protein interactions, and other protein characteristics (posttranslation modification) will help to shorten the RISC component candidate list.

20. Post-MS RISC component candidate tests are necessary to identify a new RISC component in both biophysical protein–-protein interaction (PPI) with known RISC component(s) and biological functions in RISC procedure(s). Nuclear magnet resonance (NMR) is a powerful and unique technique for the study of PPIs, particularly in the study of crucial weak PPIs (Fig. 2). There are many other useful methods to study PPIs, including GST-pull down assay, isothermal titration calorimetry assay, in cell colocalization assay, and so on.

References

1. Ota H, Sakurai M, Gupta R, Valente L, Wulff BE, Ariyoshi K, Iizasa H, Davuluri RV, Nishikura K (2013) ADAR1 forms a complex with Dicer to promote microRNA processing and RNA-induced gene silencing. Cell 153 (3):575–589. https://doi.org/10.1016/j.cell.2013.03.024
2. Yoda M, Cifuentes D, Izumi N, Sakaguchi Y, Suzuki T, Giraldez AJ, Tomari Y (2013) Poly (A)-specific ribonuclease mediates 3′-end trimming of Argonaute2-cleaved precursor microRNAs. Cell Rep 5(3):715–726. https://doi.org/10.1016/j.celrep.2013.09.029
3. Roberts TC (2015) The microRNA Machinery. Adv Exp Med Biol 887:15–30. https://doi.org/10.1007/978-3-319-22380-3_2
4. Miyoshi T, Takeuchi A, Siomi H, Siomi MC (2010) A direct role for Hsp90 in pre-RISC formation in Drosophila. Nat Struct Mol Biol 17(8):1024–1026. https://doi.org/10.1038/nsmb.1875
5. Ye X, Huang N, Liu Y, Paroo Z, Huerta C, Li P, Chen S, Liu Q, Zhang H (2011) Structure of C3PO and mechanism of human RISC activation. Nat Struct Mol Biol 18 (6):650–657. https://doi.org/10.1038/nsmb.2032
6. Yi T, Arthanari H, Akabayov B, Song H, Papadopoulos E, Qi HH, Jedrychowski M, Guttler T, Guo C, Luna RE, Gygi SP, Huang SA, Wagner G (2015) eIF1A augments Ago2-mediated Dicer-independent miRNA biogenesis and RNA interference. Nat Commun 6:7194. https://doi.org/10.1038/ncomms8194
7. Bhisutthibhan J, Meshnick SR (2001) Immunoprecipitation of [(3)H]dihydroartemisinin translationally controlled tumor protein (TCTP) adducts from Plasmodium falciparum-infected erythrocytes by using anti-TCTP antibodies. Antimicrob Agents Chemother 45(8):2397–2399. https://doi.org/10.1128/AAC.45.8.2397-2399.2001
8. Lee HW, Ryu JY, Yoo J, Choi B, Kim K, Yoon TY (2013) Real-time single-molecule coimmunoprecipitation of weak protein-protein interactions. Nat Protoc 8(10):2045–2060. https://doi.org/10.1038/nprot.2013.116
9. Shevchenko A, Tomas H, Havlis J, Olsen JV, Mann M (2006) In-gel digestion for mass spectrometric characterization of proteins and proteomes. Nat Protoc 1(6):2856–2860. https://doi.org/10.1038/nprot.2006.468
10. Gygi SP, Han DK, Gingras AC, Sonenberg N, Aebersold R (1999) Protein analysis by mass spectrometry and sequence database searching: tools for cancer research in the post-genomic era. Electrophoresis 20(2):310–319. https://doi.org/10.1002/(SICI)1522-2683(19990201)20:2<310::AID-ELPS310>3.0.CO;2-M
11. McAllister FE, Niepel M, Haas W, Huttlin E, Sorger PK, Gygi SP (2013) Mass spectrometry based method to increase throughput for kinome analyses using ATP probes. Anal

Chem 85(9):4666–4674. https://doi.org/10.1021/ac303478g

12. Villen J, Gygi SP (2008) The SCX/IMAC enrichment approach for global phosphorylation analysis by mass spectrometry. Nat Protoc 3(10):1630–1638. https://doi.org/10.1038/nprot.2008.150
13. Wuthrich K, Wagner G (1975) NMR investigations of the dynamics of the aromatic amino acid residues in the basic pancreatic trypsin inhibitor. FEBS Lett 50(2):265–268
14. Takeuchi K, Wagner G (2006) NMR studies of protein interactions. Curr Opin Struct Biol 16(1):109–117. https://doi.org/10.1016/j.sbi.2006.01.006
15. Marintchev A, Frueh D, Wagner G (2007) NMR methods for studying protein-protein interactions involved in translation initiation. Methods Enzymol 430:283–331. https://doi.org/10.1016/S0076-6879(07)30012-8

Chapter 12

Using Tet-Off Cells and RNAi Knockdown to Assay mRNA Decay

Thomas D. Baird and J. Robert Hogg

Abstract

Cellular mRNA levels are determined by the competing forces of transcription and decay. A wide array of cellular mRNA decay pathways carry out RNA turnover either on a constitutive basis or in response to changing cellular conditions. Here, we outline a method to investigate mRNA decay that employs RNAi knockdown of known or putative decay factors in commercially available Tet-off cell systems. Reporter mRNAs of interest are expressed under the control of a tetracycline-regulated promoter, allowing pulse-chase mRNA decay assays to be conducted. Levels of reporter and constitutively expressed control RNAs throughout the decay assay time course are detected by traditional northern blot analysis and used to calculate mRNA half-lives. We describe the utility of this approach to study nonsense-mediated mRNA decay substrates and factors, but it can be readily adapted to investigate key mechanistic features that dictate the specificity and functions of any mRNA decay pathway.

Key words mRNA decay, Nonsense-mediated mRNA decay (NMD), RNA stability, UPF1, Tet-off cells, RNA half-life, 3′ untranslated region

1 Introduction

mRNA decay is a highly dynamic process central to the regulation of gene expression and maintenance of cellular homeostasis [1, 2]. A major contributor to this regulation is the nonsense-mediated mRNA decay (NMD) pathway, which degrades diverse mRNAs in all eukaryotes. In addition to serving as a quality control pathway for aberrant transcripts containing premature termination codons (PTCs) resulting from genetic mutations or errors during mRNA biogenesis, the NMD pathway degrades 5–10% of apparently normal human mRNAs [3].

To investigate the activities and regulation of the mammalian NMD pathway, we describe an approach coupling RNAi-mediated protein depletion with a tetracycline-regulated reporter mRNA decay assay in human cells (Fig. 1). In this assay, siRNAs targeting the essential NMD factor UPF1 are reverse-transfected into HeLa

Shireen R. Lamandé (ed.), *mRNA Decay: Methods and Protocols*, Methods in Molecular Biology, vol. 1720,
https://doi.org/10.1007/978-1-4939-7540-2_12,

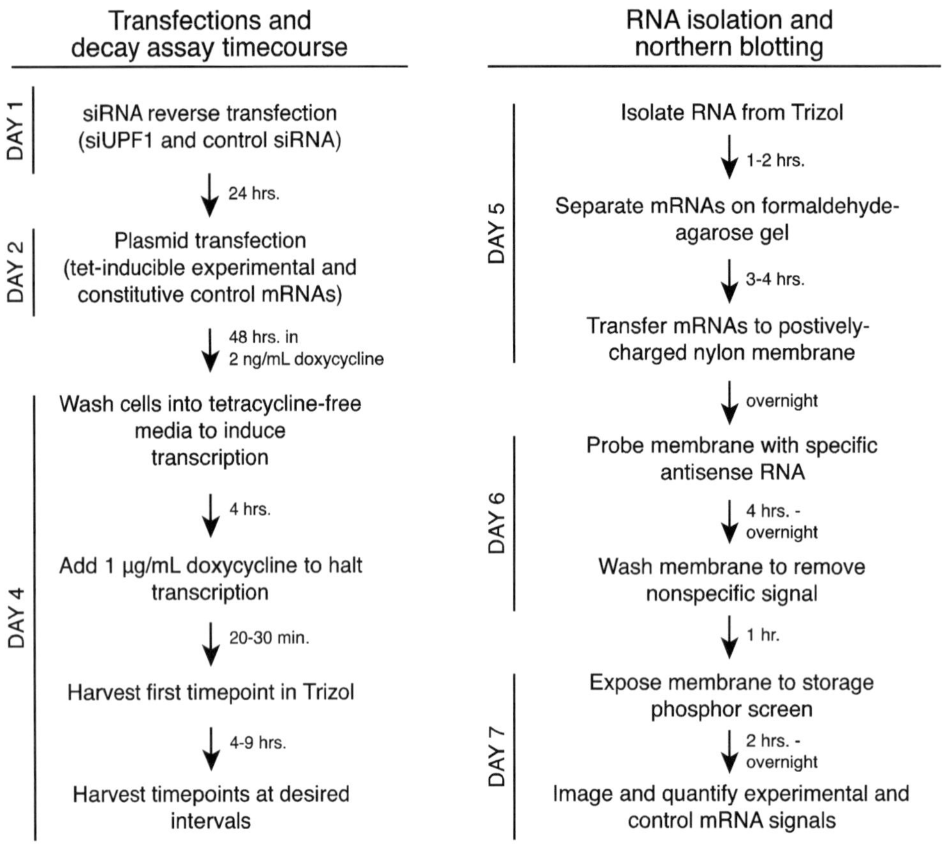

Fig. 1 Schematic of the workflow described to quantify mRNA decay using RNAi knockdown in Tet-off cell lines

cells engineered to express a Tet repressor-transcriptional activator fusion protein, rendering the NMD surveillance pathway inoperative and stabilizing NMD mRNA target substrates. Following transfection with siRNAs, plasmid DNAs encoding tetracycline-regulated reporter mRNAs and constitutively expressed transfection control mRNAs are chemically transfected into the cells.

As the reporter RNA is transcriptionally repressed in the presence of tetracycline ("Tet-off"), cells are initially grown in media supplemented with low concentrations of tetracycline or the more stable analog doxycycline. On the day of sample collection, the cells are washed with 1× PBS and supplemented with tetracycline-free media to allow a brief pulse of transcription (typically 4 h). Immediately following the induction period, tetracycline is supplemented back into the culture media to halt mRNA synthesis, and RNA samples are subsequently harvested over a range of time points. This "pulse-chase" approach allows the relative levels of the reporter RNAs to be quantified by traditional northern blot assays, using the constitutively expressed control mRNA as a normalization control. RNA half-lives can then be calculated on the basis of the ratio of these values over time, and the comparison of decay kinetics between nontargeting siRNA and siUPF1 treatments provides insight into any potential role of the NMD pathway on reporter RNA stability.

To illustrate, we describe the use of dual-fluorescent GFP-mCherry fusion reporter constructs containing either an efficient termination codon (UAA) or no termination codon (CAA) between the GFP and mCherry open reading frames. The mRNAs expressing a full-length GFP-mCherry fusion protein are highly stable, while mRNAs containing a termination codon following the GFP ORF are degraded by NMD as a result of the extended mCherry-containing 3′-UTR [4, 5]. The procedures described here are equally applicable to other commonly used NMD reporters, such as those derived from the β-globin gene [6].

While we specifically focus on NMD, this technique can easily be adapted to investigate alternative RNA decay pathways. In fact, two major utilities of the described method are the ability to both determine whether an RNA substrate (i.e., the tetracycline-regulated reporter) is subject to degradation through a specific pathway, and to investigate whether the protein depleted by RNAi is a novel regulator of a characterized RNA decay process.

2 Materials

1. Tetracycline-regulated plasmids expressing experimental mRNAs (pcTET2 2FP) and plasmids containing the CMV promoter for constitutive expression of control mRNAs (pcGFP-bGH).
2. Tetracycline-regulated Tet-off cell lines (Clontech).
3. Complete DMEM media for standard cell propagation: high-glucose Dulbecco's Modified Eagle Medium, 1% penicillin/streptomycin (pen/strep), 2.5 mM L-glutamine, 10% heat-inactivated fetal bovine serum.
4. Complete DMEM media for siRNA reverse transfection: high-glucose Dulbecco's Modified Eagle Medium, 1% pen/strep, 2.5 mM L-glutamine, 20% heat-inactivated fetal bovine serum.
5. Complete DMEM media for transcriptional induction: high-glucose Dulbecco's Modified Eagle Medium, 1% pen/strep, 2.5 mM L-glutamine, 10% Tet System Approved fetal bovine serum (Clontech).
6. Opti-MEM® reduced serum media (Gibco).
7. Sterile 1× PBS: phosphate buffered saline solution, pH 7.4.
8. siRNA (20 μM) targeting gene of interest and a nontargeting control.

9. Lipofectamine RNAiMAX® transfection reagent (ThermoFisher Scientific) or your chemical transfection reagent of choice for delivery of siRNA.
10. TurboFect® transfection reagent (ThermoFisher Scientific) or your chemical transfection reagent of choice for delivery of reporter plasmid DNA.
11. Doxycycline hyclate: 1 mg/mL stock solution dissolved in ultrapure H_2O and sterile-filtered.
12. Tissue culture plates, 24 well, flat bottom with low evaporation lid.
13. TRIzol® reagent (Ambion/ThermoFisher Scientific).
14. Sterile, RNase-free microfuge tubes.
15. Parafilm M laboratory film (Bemis).
16. Molecular-grade chloroform.
17. Molecular-grade isopropanol.
18. Molecular-grade ethanol.
19. GlycoBlue™ Coprecipitant (15 mg/mL) (ThermoFisher Scientific).
20. Formamide gel loading buffer: Deionized formamide, 15 mM EDTA, 0.1% Bromophenol blue, 0.15% Xylene cyanol.
21. Molecular-grade 37% formaldehyde solution.
22. 10× MOPS gel running buffer: 0.2 M MOPS, 40 mM sodium acetate, 5 mM EDTA, pH 7.0.
23. Horizontal gel electrophoresis system with power supply.
24. 2× MOPS/formaldehyde gel loading buffer.
25. 20× SSC: 3 M sodium citrate, 0.3 M sodium chloride, pH 7.0.
26. T7 in vitro transcription kit: including buffer, dNTPs, and T7 RNA polymerase.
27. α^{32}P-UTP (10 mCi/mL).
28. SUPERase-In™ RNase Inhibitor (ThermoFisher Scientific).
29. illustra™ Probe Quant™ G-50 micro columns (GE).
30. Geiger counter.
31. Whatman™ 3MM chromatography paper.
32. Amersham Hybond™-XL blotting membrane (GE).
33. 265 nm UV Crosslinker (e.g., Spectrolinker XL-1500 UV Crosslinker, Spectronics Corporation).
34. Hybridization oven mesh sheets (e.g., HYBAID, ThermoFisher Scientific).
35. Hybridization bottles (e.g., HYBAID, ThermoFisher Scientific).

36. ULTRAhyb™ Ultrasensitive Hybridization Buffer (Thermo-Fisher Scientific).
37. Hybridization Oven (e.g., HB-1000 Hybridizer, UVP).
38. Typhoon Trio Plus variable mode imager (GE), or comparable scanner for sensitive quantitation of ^{32}P radioisotope.
39. ImageQuant (GE), or comparable imaging software for quantification of northern blot signals.

3 Methods

3.1 siRNA Treatment and Reporter Transfection

All cell culture work should be performed in a sterile tissue culture hood using aseptic technique. Cells are grown throughout the duration of the experiment at 37 °C and 5% CO_2 in DMEM with 10% FBS and 1% pen/strep.

Day 1

1. A reverse transfection is performed to deplete cells of endogenous gene expression. Pipette 1.5 μL (stock concentration = 20 μM) nontargeting or UPF1-specific siRNA in each well of a 24-well plate (*see* **Note 1**).
2. Make a transfection master mix consisting of 250 μL Opti-MEM® and 1.5 μL RNAiMAX® per well. Account for technical replicates and pipetting error in calculating the final volume for each sample.
3. Add the above transfection mix to each well, thoroughly mixing by gently rocking the plate back and forth a few times. Allow the siRNA, OptiMEM®, and RNAiMAX® to complex at room temperature for 30–40 min.
4. During the incubation (about 10 min before the end), prepare the HeLa Tet-off cells for plating (*see* **Note 2**). Remove the normal growth media (10% FBS DMEM), rinse with 1× PBS, and trypsinize the cells thoroughly. Resuspend the cells in DMEM with 20% FBS and count (*see* **Note 3**). Add 2×10^4 cells in 250 μL DMEM with 20% FBS to each well. Rock the plate gently to distribute evenly and incubate the cells with the siRNA and transfection reagent for 10 min at room temperature. After the 10 min incubation, return plates to the cell culture incubator.

Day 2

5. In a tissue culture hood, co-transfect cells with plasmid DNAs expressing the experimental tetracycline-regulated mRNAs (pcTET2 2FP) and a constitutively expressed control (pcGFP-bGH) (*see* **Note 4**). For consistency in delivery, make a transfection mix consisting of 100 ng experimental pcTET2

2FP DNA, 25 ng control pcGFP-bGH DNA, 75 ng empty vector (e.g., pcDNA3.1), 100 μL OptiMEM, and 0.4 μL TurboFect per well. Calculate the master mix to include all of the time points being tested, as well as additional samples to allow for loss from pipetting error. It is important to add gently vortexed TurboFect to the transfection mix as the last component.

6. Incubate the transfection mix for 15–20 min at room temperature in the tissue culture hood.
7. During the incubation process, dilute a 1 mg/mL freshly made doxycycline stock 1:1000 in OptiMEM.
8. Calculate the total volume required for a final concentration of 2 ng/mL doxycycline in each well and add this to the transfection mix (*see* **Note 5**).
9. At the end of the incubation period, carefully add the transfection mix dropwise to each well, rock the plate gently to mix, and return the plates to the cell culture incubator.

Day 3

10. Inspect cells under a light microscope to verify even distribution and healthy morphology. Fresh media and doxycycline may be added at this step, though this is not required for robust transcriptional repression and may lead to loss of cells in lines that adhere poorly.

Day 4

11. In a tissue culture hood, aspirate off the media, gently wash each well with 0.5 mL 1× PBS, and replace media with 0.5 mL DMEM + 10% tetracyline-free FBS (*see* **Note 6**). Incubate the cells at 37 °C for 4 h in tetracycline-free media to de-repress transcription of reporter mRNAs (pcTET2 2FP).
12. After the 4 h incubation period, supplement each well with doxycycline at a final concentration of 1 μg/mL (i.e., dilute 1 mg/mL doxycycline stock 1:10 in OptiMEM and add 5 μL to each well).
13. Start the experimental time course 20–30 min following the addition of doxycycline to provide sufficient time to repress all transcription of the reporter mRNAs.
14. Collect time points at 0, 3, 6, and 9 h by harvesting the cells in the hood with 500 μL TRIzol/well (*see* **Note** 7). Following the TRIzol harvest, rinse empty wells twice with 1 mL 1× PBS to remove residual TRIzol from harvested wells and return to incubator for later time points.

15. Store TRIzol samples in RNase-free microfuge tubes at −20 °C for short term storage (e.g., 2 weeks) or −80 °C for longer periods of time.

3.2 RNA Extraction and Purification of TRIzol Samples

Before proceeding to the RNA isolation step, chill a tabletop microcentrifuge to 4 °C and label new microfuge tubes in advance.

1. Following a slightly adapted version of the manufacturer's protocol, add 100 μL of chloroform to the 500 μL thawed TRIzol sample and cap the tube securely. Shake the tube vigorously by hand for 15 s, and incubate at room temperature (25 °C) for 2–3 min.
2. Centrifuge the samples at 12,000 × *g* for 15 min at 4 °C.
3. Remove the aqueous phase of the sample by angling the tube at 45° and pipetting the solution into a new labeled tube (*see* **Note 8**). Importantly, avoid drawing any of the interphase or organic layer into the pipette when removing the aqueous phase. Dispose of the organic phenol phase according to your institution's waste disposal guidelines.
4. Add 250 μL of 100% isopropanol to the aqueous phase (per 500 μL TRIzol used for homogenization). When precipitating RNA from small sample quantities, add 2 μL GlycoBlue™ coprecipitant as a carrier. This step increases both pellet mass and visibility.
5. Incubate at room temperature for 10 min.
6. Centrifuge samples at 12,000 × *g* for 10 min at 4 °C. The RNA should be visible following centrifugation as a small gel-like blue pellet on the side and bottom of the tube. During subsequent steps, carefully monitor the presence of this pellet when cautiously removing supernatant.
7. Carefully pipette the supernatant from the tube, leaving only the RNA pellet.
8. Wash the pellet with 500 μL of 75% ethanol (per 500 μL TRIzol used in the initial homogenization). Vortex the sample briefly, then centrifuge the tube at 7500 × *g* for 5 min at 4 °C.
9. Carefully pipette the wash supernatant from the tube, leaving only the RNA pellet, and centrifuge the tube again at 7500 × *g* for 1 min at 4 °C to allow removal of residual ethanol (*see* **Note 9**).
10. Air-dry the pellet for 3–5 min at room temperature. Do not allow the RNA to dry completely, as this can result in a decrease in pellet solubility.
11. Resuspend RNA pellet by adding 10 μL Formamide LB to each sample. Shake samples at 50 °C for 5–10 min, which can then either be stored at −80 °C for future detection or used immediately for northern blotting.

3.3 Detection and Quantification of Reporter mRNAs by Northern Blot

3.3.1 Preparing, Running, and Transferring the Gel

1. The evening before running the gel, prepare 2 L of 1× MOPS gel running buffer and refrigerate at 4 °C.
2. Prepare a 1.2% agarose/formaldehyde gel. Combine the appropriate amount of agarose with ultrapure H_2O in a glass Erlenmeyer flask and microwave in 15 s intervals to melt. Once all of the agarose is in solution, carefully swirl the contents and place the flask in a 60 °C H_2O bath to equilibrate the temperature.
3. After equilibrating the agarose solution to 60 °C, add 10× MOPS at 12% of the volume of H_2O used to dissolve the agarose (e.g., if 100 mL of H_2O was used to dissolve agarose, add 12 mL of 10× MOPS) (*see* **Note 10**).
4. Add 37% formaldehyde at 7% of the volume of H_2O used to dissolve the agarose (e.g., if 100 mL of H_2O was used to dissolve agarose, add 7 mL of 37% formaldehyde). Swirl the solution and cast the agarose/formaldehyde gels in a chemical hood (*see* **Note 11**).
5. To further prepare the RNA samples for electrophoresis, mix the RNA samples in formamide LB 1:1 with 10 μL 2× MOPS/formaldehyde LB. Heat the samples at 80 °C for 2–5 min (*see* **Note 12**).
6. While the samples are incubating, add 50 mL formaldehyde to the 2 L of cold 1× MOPS buffer. Gently pour this solution over the agarose/formaldehyde gel in the electrophoresis chamber (*see* **Note 13**).
7. Before loading the samples, it is important to clean each well by pipetting 1× MOPS/formaldehyde buffer in and out of each well, preventing band smearing. Once the wells are all rinsed, load 20 μL of each sample into each well in the desired order for the final northern blot image. Run the gel at 5 V/cm for approximately 3 h or until the two dye fronts have adequately separated.
8. Once the run is complete, begin the gel transfer by first trimming off the elevated lips bordering the gel without cutting into sample lanes. This will allow the gel to lay as flat as possible during the transfer. In an appropriately sized Pyrex (or comparable) glass dish, rinse the gel on a rotator 3× in ultrapure H_2O.
9. After the third wash, replace the water with 10× SSC (diluted from 20× SSC with ultrapure H_2O), and slowly agitate on rotator for 30–60 min.
10. While the gel is equilibrating in 10× SSC, place a glass plate over a large Pyrex (or comparable) glass tray filled with 600–800 mL of 10× SSC. Place two pieces of 3MM chromatography paper over the glass plate in such a way they dip into

the 10× SSC buffer. Wet the paper with 10× SSC, and use a 25 mL plastic pipette to gently roll out any puddles, bubbles, or wrinkles before proceeding to the next step.

11. Place the equilibrated gel upside down on top of the 3MM chromatography paper, again ensuring that no bubbles or wrinkles are introduced. Using the gel as a template for size, cut a piece of Amersham Hybond™-XL blotting membrane of comparable size, situate on top of the gel, and wet the membrane with the 10× SSC buffer clearing bubbles by rolling with the plastic pipette.
12. Wet a piece of 3MM chromatography paper with the 10× SSC and place on top of the membrane (*see* **Note 14**). Roll this with the pipette to remove any bubbles, and repeat with two additional pieces of 3MM paper (total of three).
13. To block buffer from transferring outside of the gel, place four pieces of Parafilm, one on each side of the gel, effectively creating a dam around the transfer sandwich. Finally, add three pieces of dry 15 × 20 cm 3MM papers on top of the stack, followed by about 10 cm worth of dry paper towels, and lastly a Pyrex tray or other flat surface weighed down with a 500 mL bottle of liquid. Leave overnight for capillary transfer.

3.3.2 Preparation of the Riboprobe

1. To prepare the radioactive riboprobe used for northern blotting, first linearize the plasmid DNA serving as the transcription template using the proper restriction enzyme. Purify the linearized DNA by phenol–chloroform extraction and ethanol precipitation or spin column method of choice, and resuspend the DNA in ultrapure H_2O to a final concentration of 0.05 μg/μL. Alternatively, the desired probe fragment can be generated by PCR, using a reverse primer containing the T7 promoter sequence.
2. Following your institution's approved radioactive materials guidelines, prepare the in vitro transcription reaction by assembling the following reaction:

   1 μL 10× T7 buffer.

   1 μL 50 mM DTT.

   1 μL 5 mM ATP, GTP, CTP, and 0.1 mM UTP.

   1 μL linearized plasmid (0.05 μg/μL).

   4.0 μL α^{32}P-UTP (10 mCi/mL).

   0.5 μL SUPERase-In™ ribonuclease inhibitor.

   0.5 μL T7 RNA polymerase.

   – μL ddH_2O to make 10 μL total.

Incubate at 37 °C for 1 h and dilute final reaction to 50 μL by adding 40 μL TE buffer.

3. To remove unincorporated dNTPs, first place a Sephadex G50 column in an open microfuge tube and centrifuge at 735 × *g* for 1 min. Transfer the spin column to a new tube and add the probe mix directly to the center of the column. Centrifuge at 735 × *g* for 2 min and transfer eluate containing the probe to a new tube. Use a Geiger or scintillation counter to ensure the probe is radioactive, and dilute the eluate to 500 μL with TE-buffer. As only 100 μL of this probe is required per each membrane hybridization, the remaining reaction can be stored in a radioactive material storage box at −20 °C.

3.3.3 Membrane Hybridization and Imaging

1. The morning after the transfer, wash the membrane gently with ultrapure H_2O to remove all salt. Set the membrane on 3MM paper and let dry at room temperature for approximately 30 min.
2. Place the membrane on top of 3MM paper with the RNA sample-side facing up, and crosslink the RNA to the membrane using a 254 nm UV crosslinker set to deliver 1200 × 100 μJ/cm^2. Wet the blot in a bath of 2× SSC to avoid background, and carefully roll the blot in a nylon hybridization mesh before inserting into a hybridization bottle. It is crucial that during the rolling incubation the RNA sample-side of the membrane is facing towards the inner cavity of the tube with the surface area completely exposed to the hybridization buffer.
3. Add 10–20 mL ULTRAhyb™ Ultrasensitive Hybridization Buffer prewarmed to 65 °C, and screw on lids without over tightening. Place tubes in the hybridization oven ensuring that tube weight is balanced equally across the rotisserie rotor (simply match an unpaired tube with one filled with H_2O of an equal mass). Incubate in a hybridization oven for at least 60 min at 68 °C.
4. Add the RNA probe directly to the hybridization buffer in the bottom of the tube, being careful to avoid exposing any of the probe to the membrane at this step.
5. Incubate the blot at 68 °C for at least 4 h to overnight.
6. Decant the hybridization solution in an appropriate radioactive liquid waste container and rinse the hybridization tube with ~30 mL 2×SSC, 0.1% SDS solution. Decant the rinse and wash the blot twice for 5 min in 2×SSC, 0.1% SDS and twice for 15 min in 0.1×SSC, 0.1% SDS.
7. Following the final wash, remove the blot from the hybridization tube with forceps or hemostats. Lay the blot on plastic wrap to briefly dry, before wrapping it in a second piece of

plastic wrap. Expose to a storage phosphor screen for a few hours to overnight to detect mRNAs.

8. Image storage phosphor screen using a Typhoon Trio Plus variable mode imager (GE), or comparable phosphorimager. Adjust exposure time as necessary to increase sensitivity to allow visualization of faint bands or decrease sensitivity to avoid saturation of the probe signal.
9. Use ImageQuant or equivalent image analysis software to quantify the signals derived from the experimental and control mRNAs at each timepoint. Normalize the experimental signal to the control signal and set the initial abundance of the experimental mRNA to 1 to calculate the fraction of experimental mRNA remaining at each timepoint. Plot the fraction mRNA remaining versus time on a semilog plot. The slope of the best-fit line can then be used to calculate the mRNA half-life, using the equation $t_{1/2} = -0.43\ln(2))/\text{slope}$.

4 Notes

1. To disrupt NMD, we use a previously characterized siRNA specific to UPF1 [7]. The use of other commercially available targeting siRNAs and nontargeting controls can be substituted as necessary.
2. Our method describes using the HeLa Tet-Off® cell line, but there are several other commercial examples of both human and murine lines that express the tetracycline-regulated Tet-Off® system. For example, the HEK-293 Tet-Off® line is useful for achieving high transfection efficiency. It is important to note, however, that subtle changes to the protocol may be required when using alternative lines. It is our experience that certain lines including HEK-293 require an additional replating step on coated cell culture plates (e.g., Pure-Coat amine) prior to transfection with the reporter mRNAs. Such modifications for enhanced cell attachment and growth may be necessary for other cell types, as well.
3. DMEM with 20% FBS (twice the normal) is required to account for the lack of serum in the OptiMEM® media.
4. The pcTET2 constructs used in this protocol were based on constructs originally introduced by Lykke-Andersen and Steitz [6] and were previously described [5]. We find these reporters to be useful as they provide a platform to study mRNA decay, translational readthrough, and mRNP composition in parallel. However, any tetracycline-regulated expression plasmid can be employed with this system to discern RNA stability and dependency on trans-acting factors. High-quality plasmid DNA that

is endotoxin-free should be used for optimal transfection efficiency.

5. We prefer to use the tetracycline analogue doxycycline in these experiments for its enhanced stability. Tetracycline may be used, but requires supplementing the media with fresh drug on a daily basis. Doxycycline stock solution should be stored at 4 °C, protected from light, and used within 7 days of preparation. For HeLa Tet-off cells, we use 2 ng/mL doxycycline to repress transcription. For different cell types, it may be necessary to titrate doxycycline to optimize the repression prior to the pulse and the magnitude and kinetics of activation following removal of doxycycline.
6. Standard fetal bovine serum often contains trace amounts of tetracycline contaminants, which can lead to undesirable background repression in the Tet-Off lines.
7. These time points capture the dynamics of decay for the pcTET2 2FP experimental mRNAs described. For other experimental constructs, an appropriate time course will need to be determined empirically.
8. The mixture separates into a lower red organic phase, an interphase, and a colorless upper aqueous phase. RNA remains exclusively in the aqueous phase. The upper aqueous phase is ~50% of the total volume.
9. This additional centrifuge step pools residual ethanol from the wash into the bottom of the tube, which can then be gently pipetted away, allowing for the pellet to air dry more quickly and efficiently.
10. To make 1 L of 10× MOPS, dissolve 41.8 g of MOPS, 3.29 g of sodium acetate, and 1.46 g of EDTA to 900 mL of ddH_2O. Adjust the pH to 7.0, and bring the final volume to 1 L. Filter sterilize the solution, protect from light by covering the bottle with aluminum foil, and store at 4 °C.
11. If the formaldehyde solution is cloudy, use a new bottle for all of the steps described in this protocol.
12. To make 2× MOPS/formaldehyde LB (prepare fresh every time), prepare a 4:1 solution of formaldehyde–10× MOPS, and use 0.5 M EDTA stock to achieve a final concentration of ~15 mM EDTA.
13. We have found it is good practice to thoroughly rinse both the casting box and combs prior to casting the gel. Allow sufficient time for combs to air dry for best well formation.
14. We suggest marking a designated membrane corner with a pencil to establish orientation for future reference.

Acknowledgments

This work was supported by the Intramural Research Program, National Institutes of Health, National Heart, Lung, and Blood Institute. We thank Zhiyun Ge, Aparna Kishor, and Stacey Baker for troubleshooting aspects of this protocol.

References

1. Chen CA, Shyu AB (2016) Emerging themes in regulation of global mRNA turnover in cis. Trends Biochem Sci. https://doi.org/10.1016/j.tibs.2016.08.014
2. Palumbo MC, Farina L, Paci P (2015) Kinetics effects and modeling of mRNA turnover. Wiley Interdiscip Rev RNA 6(3):327–336. https://doi.org/10.1002/wrna.1277
3. Karousis ED, Nasif S, Muhlemann O (2016) Nonsense-mediated mRNA decay: novel mechanistic insights and biological impact. Wiley Interdiscip Rev RNA 7(5):661–682. https://doi.org/10.1002/wrna.1357
4. Baker SL, Hogg JR (2017) A system for coordinated analysis of translational readthrough and nonsense-mediated mRNA decay. PLoS One 12 (3):e0173980. https://doi.org/10.1371/journal.pone.0173980
5. Tang X, Zhu Y, Baker SL, Bowler MW, Chen BJ, Chen C, Hogg JR, Goff SP, Song H (2016) Structural basis of suppression of host translation termination by Moloney Murine Leukemia Virus. Nat Commun 7:12070. https://doi.org/10.1038/ncomms12070
6. Lykke-Andersen J, Shu MD, Steitz JA (2000) Human Upf proteins target an mRNA for nonsense-mediated decay when bound downstream of a termination codon. Cell 103 (7):1121–1131
7. Mendell JT, ap Rhys CM, Dietz HC (2002) Separable roles for rent1/hUpf1 in altered splicing and decay of nonsense transcripts. Science 298(5592):419–422. https://doi.org/10.1126/science.1074428

Chapter 13

Identifying Cellular Nonsense-Mediated mRNA Decay (NMD) Targets: Immunoprecipitation of Phosphorylated UPF1 Followed by RNA Sequencing (p-UPF1 RIP–Seq)

Tatsuaki Kurosaki, Mainul Hoque, and Lynne E. Maquat

Abstract

Recent progress in the technology of transcriptome-wide high-throughput sequencing has revealed that nonsense-mediated mRNA decay (NMD) targets ~10% of physiologic transcripts for the purpose of tuning gene expression in response to various environmental conditions. Regardless of the eukaryote studied, NMD requires the ATP-dependent RNA helicase upframeshift 1 (UPF1). It was initially thought that cellular NMD targets could be defined by their binding to steady-state UPF1, which is largely hypophosphorylated. However, the propensity for steady-state UPF1 to bind RNA nonspecifically, coupled with regulated phosphorylation of UPF1 on an NMD target serving as the trigger for NMD, made it clear that it is phosphorylated UPF1 (p-UPF1), rather than steady-state UPF1, that can be used to distinguish cellular NMD targets from cellular RNAs that are not. Here, we describe the immunoprecipitation of p-UPF1 followed by RNA sequencing (p-UPF1 RIP–seq) as a transcriptome-wide approach to define physiologic NMD targets.

Key words mRNP, RNA helicase, Phosphorylated protein, p-UPF1, Immunoprecipitation, Deep sequencing, RIP–seq

1 Introduction

Nonsense-mediated mRNA decay (NMD) is a quality-control mechanism that degrades aberrant mRNAs harboring a premature termination codon (PTC), many of which are caused by mistakes made during pre-mRNA splicing [1, 2]. NMD also downregulates ~10% of physiologic transcripts, which may harbor an upstream open reading frame (uORF), an intron sufficiently downstream of a normal termination codon, a UGA selenocysteine codon, or a long 3′-untranslated region (3′UTR) [3–5]. The central NMD factor is the ATP-dependent RNA helicase upframeshift 1 (UPF1). High-throughput sequencing, which is a powerful tool to define protein-binding sites on RNA, revealed that steady-state UPF1, which is largely hypophosphorylated, promiscuously binds

Shireen R. Lamandé (ed.), *mRNA Decay: Methods and Protocols*, Methods in Molecular Biology, vol. 1720,
https://doi.org/10.1007/978-1-4939-7540-2_13, © Springer Science+Business Media, LLC 2018

physically accessible cellular RNAs with relatively low affinity [6–11]. This low level of translation-independent, nonspecific binding confounds the ability to UPF1 binding to identify cellular NMD targets [6, 11].

Instead, cellular NMD targets can be identified by localizing the RNA-binding sites of the activated form of UPF1, which is generated in a translation-dependent regulated series of choreographed steps that result in UPF1 phosphorylation by the phosphatidylinositol 3-kinase-related protein kinase, suppressor with morphogenic effect on genitalia 1 (SMG1). Phosphorylated UPF1 (p-UPF1) not only represses further rounds of translation initiation on the NMD target [12] but also indirectly recruits exonucleolytic activities to the NMD target via interactions with SMG5 and SMG7 and directly recruits the endonucleolytic activity SMG6 [13–17]. Furthermore, p-UPF1 binds NMD target 3′UTRs [6–11]. Protein phosphatase 2A (PP2A) dephosphorylates UPF1, leading to UPF1 recycling and most likely also the prevention of non-NMD target decay [15, 18–20].

Since the level of p-UPF1 constitutes only a small fraction of the bulk of cellular UPF1, it is barely detectable by western blotting using an antibody to steady-state UPF1 [15], and RNA immunoprecipitation (IP) using an antibody to p-UPF1 is generally inadequate to generate sufficient RNA for cDNA construction and high-throughput sequencing. Thus, to map p-UPF1 binding sites on cellular RNAs, it is important to increase the steady-state level of p-UPF1 using one of several strategies. The first strategy involves transfecting cells with either siRNA or shRNA to downregulate cellular PP2A, SMG5, SMG6, or SMG7, each of which is directly or indirectly involved in dephosphorylating p-UPF1 [14, 15, 18–21]. The second strategy involves transfecting cells with plasmid DNA encoding a tagged variant of UPF1 that harbors one or more mutations in the UPF1 helicase or ATPase domain that increase the level of p-UPF1 ~ 4–5-fold above its normal level [15, 21]. The third strategy exposes cells to a protein phosphatase inhibitor such as okadaic acid to increase the cellular level of p-UPF1 [12, 15, 22]. Since the first two approaches employ transfections of siRNA, shRNA, or plasmid DNA, careful experimental optimization is required. Incubating cells with okadaic acid, in contrast, is fast (involving an incubation period of $\leq$3 h), and is applicable to most if not all cultured cell types. Moreover, since okadaic acid inhibits NMD [12], it also increases the fraction of sequence reads that derive from cellular NMD targets.

In this chapter, we provide a schematic overview (Fig. 1) and detailed protocol for immunoprecipitating p-UPF1 from okadaic acid-treated human embryonic kidney (HEK) 293T cells for the purpose of identifying cellular NMD targets. We call this method p-UPF1 RIP—seq. This protocol is generally applicable to the many cultured cell types that are available from many mammalian species.

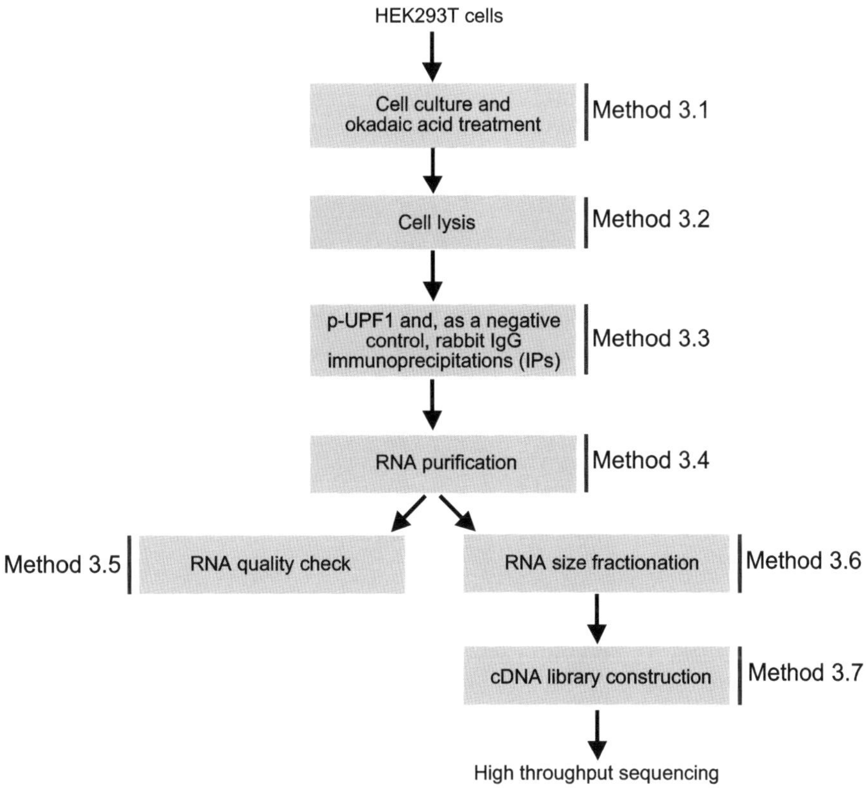

Fig. 1 Schematic for p-UPF1 RIP-Seq

2 Materials

2.1 Cell Culture and Okadaic Acid Treatment

1. Dulbecco's Modified Eagle's Medium (DMEM) supplemented with 10% fetal bovine serum (FBS) (Gibco | Thermo Fisher Scientific).
2. Okadaic acid: 100 μM okadaic acid (Sigma-Aldrich) in dimethyl sulfoxide (DMSO) (*see* **Note 1**).
3. Ice-cold phosphate-buffered saline (PBS). For this and all other solutions, water is deionized (Aries FilterWorks).
4. Corning Cell Lifter (Sigma-Aldrich).

2.2 Cell Lysis

1. Hypotonic Gentle Lysis Buffer: 10 mM Tris–HCl (pH 7.4), 10 mM NaCl, 10 mM EDTA, 0.5% (w/w) Triton X-100, 1× of a 7× Roche cOmplete Mini EDTA-free Protease Inhibitor Cocktail (Sigma-Aldrich), 1× of a 10× Roche PhosSTOP Phosphatase Inhibitor Cocktail (Sigma-Aldrich) (*see* **Note 2**).
2. 5 M NaCl.

2.3 p-UPF1 Immunoprecipitation

1. Antibodies: Anti-phospho-UPF1 Ser1116 antibody (anti-phospho-UPF1 (Ser1127), EMD Millipore); Control rabbit IgG (Sigma-Aldrich) (*see* **Note 3**).
2. Dynabeads Protein A (Thermo Fisher Scientific).
3. Ice-cold NET-2 Buffer (0.1% Nonidet P-40 (NP-40)): 50 mM Tris–HCl (pH 7.4), 150 mM NaCl, 0.1% NP-40.
4. Ice-cold NET-2 Buffer (0.5% NP-40): 50 mM Tris–HCl (pH 7.4), 150 mM NaCl, 0.5% NP-40.
5. High Salt Wash Buffer: 20 mM Tris–HCl (pH 7.4), 325 mM NaCl, 2 mM EDTA, 1% Triton X-100, 0.1% SDS.
6. Ambion RNase I (100 U/μl) (Thermo Fisher Scientific).
7. 10× RNase buffer: 200 mM Tris–HCl (pH 7.5), 1 M NaCl, 1 mM EDTA, 0.1% Triton X-100.
8. 2× SDS-PAGE Sample Elution Buffer: 125 mM Tris–HCl (pH 6.8), 4% SDS, 20% glycerol, 0.1% bromophenol blue, 10% 2-mercaptoethanol (*see* **Note 4**).
9. Magnetic Tube Rack (New England Biolabs or Bio Rad Laboratories).
10. Tube Rotator (Thermo Fisher Scientific).

2.4 RNA Preparation for cDNA Library Construction

1. T4 Polynucleotide Kinase (10,000 U/ml) (New England Biolabs).
2. ATP, [γ^{32}P] 3000 Ci/mmol (PerkinElmer).
3. Invitrogen TRIzol Reagent (Thermo Fisher Scientific).
4. Chloroform.
5. DynaMarker Prestain Marker for Small RNA (Diagnicine).
6. RNA Extraction Buffer: 20 mM Tris–HCl (pH 7.5), 300 mM sodium acetate, 2 mM EDTA, 0.2% (v/v) SDS.
7. Corning Costar Spin-X Centrifuge Tube Filter (pore size: 0.45 μm) (Sigma-Aldrich).
8. Ethanol (100% and 75%).
9. 2-Propanol.
10. Glycogen Solution (20 mg/ml) (Thermo Fisher Scientific).
11. NanoDrop 1000 Spectrophotometer (Thermo Fisher Scientific).
12. Gel Loading Buffer II (Thermo Fisher Scientific).
13. 15% Polyacrylamide Slab Gel 9 (15 × 17 cm) containing 6 M Urea.
14. SYBR Gold Nucleic Acid Gel Stain (Thermo Fisher Scientific).
15. Typhoon FLA 9500 Laser Scanner or equivalent (GE Healthcare Life Sciences).

16. Molecular Imager Gel Doc XR+ System with Image Lab Software (Bio Rad Laboratories).

2.5 cDNA Library Construction for p-UPF1 RIP–seq

1. Shrimp Alkaline Phosphatase (1000 U/ml) (New England Biolabs).
2. T4 Polynucleotide Kinase (10,000 U/ml) (New England Biolabs).
3. miRNeasy Mini Kit (Qiagen).
4. T4 RNA Ligase 2, truncated (200,000 U/ml) (New England Biolabs).
5. T4 RNA Ligase 1 (10,000 U/ml) (New England Biolabs).
6. Invitrogen Superscript III Reverse Transcriptase (Thermo Fisher Scientific).
7. Phusion High-Fidelity DNA Polymerase (New England Biolabs).
8. TE Buffer: 10 mM Tris-HCl (pH 8.0), 1 mM EDTA.
9. 3′ DNA Adaptor 5′-rApp/NNNN TGGAATTCTCGGGT GCCAAGG/ddC-3′ (where rApp represents adenylation, and ddC represents dideoxycytosine) (*see* **Note 5**).
10. 5′ RNA Adaptor: 5′-GUUCAGAGUUCUACAGUCCGAC GAUCNNNN-3′ (*see* **Note 5**).
11. Reverse Transcription (RT) Primer: 5′-CCTTGGCACCC GAGAATTCCA-3′.
12. PCR Forward Primer: 5′-AATGATACGGCGACCACC GAGATCT ACACGTTCAGAGTTCTACAGTCCGA-3′.
13. PCR Reverse Primer 5′-CAAGCAGAAGA CGGCATACGA-GAT[CGTGAT]GTGACTGGAGTTCCTTGGCACCCGA-GAATTCCA-3′ (where the index region is bracketed).
14. Sequencing Primer: 5′-CTACACGTTCAGAGTTCTAC AGTCCGACGATC-3′.
15. Phenol (pH 8.0) (Sigma-Aldrich).
16. 2100 Bioanalyzer (Agilent Technologies).
17. HiSeq 2500 System (Illumina).

3 Methods

3.1 Cell Culture and Okadaic Acid Treatment

1. Add 1/500 volume of 100 μM okadaic acid to the culture medium so the final concentration is 200 nM (*see* **Note 1**) (Fig. 2).
2. Incubate at 37 °C for 3 h in a CO_2 incubator.

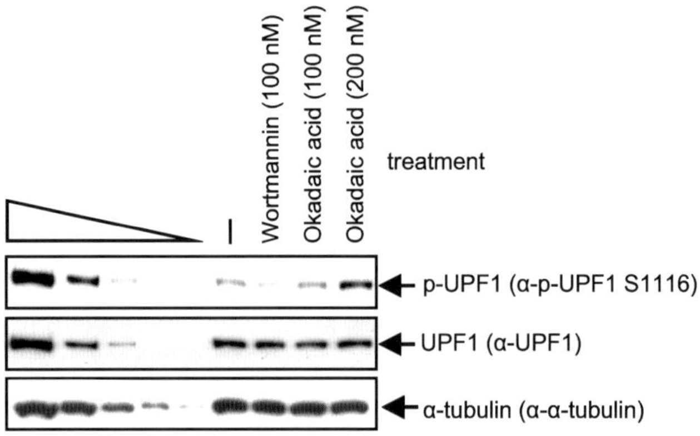

Fig. 2 Western blot of lysates of HEK293T cells (5×10^7/150-mm dish) that were untreated (–) or treated for 3 h with 100 nM wortmannin (to inhibit phosphatidyl inositol 3-kinases, including SMG1), or 100 nM okadaic acid or 200 nM okadaic acid (to inhibit type 1 and type 2A protein phosphatases). Note that 200 nM okadaic acid resulted in an ~4-fold increase in cellular p-UPF1 abundance

3. Pour off and discard medium, collect cells using a cell lifter, and pellet cells in a 1.5 ml microfuge tube at 1000 × *g* for 5 min. Wash cell pellets two times using 3 ml of ice-cold PBS.
4. Remove PBS and collect cell pellets (*see* **Note 6**).

3.2 Cell Lysis

1. Resuspend cell pellets (*see* **Note 7**) in Hypotonic Gentle Lysis Buffer and keep on ice for 10 min.
2. Mix well using a vortex mixer for 30 s. Add 5 M NaCl to a final concentration of 150 mM, and vortex vigorously for 30 s.
3. Pellet cellular debris at 15,000 × *g* and 4 °C for 10 min, and transfer the supernatant to a new 1.5 ml microcentrifuge tube (*see* **Note 8**).

3.3 p-UPF1/Rabbit IgG Immunoprecipitation (IP), and On-Bead Limited Digestion Using RNase I

1. For each cell lysate, prewash 50 μl of Dynabeads Protein A in Hypotonic Gentle Lysis Buffer containing 150 mM NaCl, concentrate the Dynabeads Protein A to the sides of the microfuge tube using a Magnetic Tube Rack on ice, and remove the wash buffer using aspiration (*see* **Note 9**).
2. Preclear each lysate by adding supernatant from Subheading 3.2, **step 3** to prewashed Dynabeads Protein A from **step 1** and mixing end-over-end using a Tube Rotator at 4 °C for 1 h.
3. While preclearing each lysate, prepare the antibody–Dynabeads Protein A (*see* **Note 10**) by adding 5–10 μg of anti-phospho-UPF1 (Ser1116) antibody or, as a negative control, rabbit IgG (*see* **Note 11**) to 50 μl of washed Dynabeads Protein A from Subheading 3.3, **step 1** in 300 μl of Hypotonic Gentle Lysis Buffer containing 150 mM NaCl (*see* **Note 12**). Rotate end-over-end at 25 °C (i.e. room temperature) for

30 min. Wash with Hypotonic Gentle Lysis Buffer containing 150 mM NaCl once more (*see* **Note 9**), and carefully remove the buffer using aspiration.

4. Centrifuge precleared lysates from **step 2** at 15,000 × *g* at 4 °C for 10 min, add the resulting supernatant to the antibody-Dynabeads Protein A mixture from **step 3**, and rotate end-over-end at 4 °C for 2 h.
5. Wash the antibody–Dynabeads Protein A mixture with 1 ml of ice-cold NET-2 Buffer (0.1% NP-40) (*see* **Note 9**). Repeat this step once more.
6. Wash the antibody–Dynabeads Protein A mixture with 300 μl of 1× RNase buffer (*see* **Note 9**).
7. Add 300 μl of 1× RNase Buffer and RNase I (100 U/μl) to 1 U/μl (*see* **Note 13**). Rotate end-over-end at 4 °C for 30 min.
8. Wash antibody–Dynabeads Protein A mixture using 1 ml of ice-cold NET-2 Buffer (0.1% NP-40) (*see* **Note 9**). Repeat this step once more.
9. Wash antibody–Dynabeads Protein A mixture with 1 ml of ice-cold NET-2 Buffer (0.5% NP-40) (*see* **Note 9**). Repeat this step once more.
10. Wash antibody–Dynabeads Protein A mixture with 1 ml of High Salt Wash Buffer (*see* **Note 9**). Repeat this step once more.
11. Carefully remove High Salt Wash Buffer, and elute RNA using 50 μl of 2× SDS-PAGE Sample Elution Buffer by incubating at 95 °C for 5 min (*see* **Note 4**).

3.4 RNA Purification

1. Transfer 30 μl of eluted IP sample into a new 1.5 ml microfuge tube and bring the volume to 100 μl with water. Add 500 μl of TRIzol Reagent and 100 μl of chloroform.
2. Rotate end-over-end for five rotations, let sit at 25 °C for 2–3 min, and resolve the aqueous and organic phases by centrifugation at 15,000 × *g* at 4 °C for 5 min.
3. Transfer upper aqueous phase to a new 1.5-ml tube, add an equal volume (~500 μl) of 2-propanol and 1 μl of Glycogen Solution, rotate end-over-end five times by hand, and pellet RNA by centrifugation at 15,000 × *g* at 4 °C for 10 min.
4. Wash RNA pellets with 1 ml of 75% ethanol, collecting the pellets by centrifugation at 15,000 × *g* for 5 min at 4 °C. Repeat this step one more time.
5. Air-dry the RNA pellets at 25 °C for 5 min, and dissolve in 15 μl of water.
6. Quantitate resuspended RNA concentration using a NanoDrop 1000 Spectrophotometer.

3.5 RNA Quality Check

1. Radiolabel 100–200 ng of RNA with 5–10 μCi of [γ^{32}P] ATP (3000 Ci/mmol) using 10 U of T4 Polynucleotide Kinase at 37 °C for 30 min.
2. Add an equal volume of Gel Loading Buffer II, and incubate at 65 °C for 5 min.
3. Load ~100 ng of radiolabeled RNA per lane onto a denaturing 15% polyacrylamide gel containing 6 M urea in parallel with 5 μl of DynaMarker Prestain Marker for Small RNA, and electrophorese at 260 V and 37 °C for 2–3 h (*see* **Note 14**).
4. Visualize the RNA using a Typhoon FLA 9500 Laser Scanner to evaluate the efficiency of RNase I digestion (*see* **Note 13**).

3.6 RNA Size Fractionation

1. Using the remaining unlabeled RNA, add an equal volume of Gel Loading Buffer II, incubate at 65 °C for 5 min, and electrophorese at 260 V for 2–3 h in a 15% polyacrylamide slab gel containing 6 M urea in parallel with the DynaMarker Prestain Marker for Small RNA (*see* **Note 14**).
2. Stain the gel with 1× SYBR Gold Nucleic Acid Gel Stain at 25 °C for 30 min, and visualize the RNA using a Molecular Imager Gel Doc XR+ System with Image Lab Software (*see* **Note 15**).
3. Using a clean razor blade, excise RNA migrating from 25 to 40 nt, pass the gel piece through a 1 ml plastic syringe (without needle) into a 1.5 ml microfuge tube, add 400 μl of RNA Extraction Buffer, and rotate end-over-end at 4 °C for 12–18 h.
4. Transfer the RNA suspension to a Corning Costar Spin-X Centrifuge Tube Filter, and remove gel debris by centrifugation at 15,000 × *g* at 4 °C for 5 min (*see* **Note 16**).
5. Purify RNA using TRIzol Reagent as described in Subheading 3.4, **steps 1–6**.

3.7 cDNA Library Construction for p-UPF1 IP-seq

1. Remove the 3′ phosphates from ~100 ng of purified RNA by incubation with 1 U of Shrimp Alkaline Phosphatase at 37 °C for 30 min, followed by enzyme inactivation by incubating at 65 °C for 10 min.
2. Phosphorylate the 5′ ends by incubation with 10 U of T4 Polynucleotide Kinase at 37 °C for 30 min, followed by enzyme inactivation by incubation at 65 °C for 20 min.
3. Purify RNA using the miRNeasy Mini Kit following Qiagen Manual instructions.
4. Ligate 1 μl of a 1 μM 3′ DNA Adaptor solution to the 3′ ends by incubating with 20 U of truncated T4 RNA Ligase 2 at 22 °C for 1 h.

5. Anneal 1 μl of a 5 μM RT primer solution to the 3′ DNA adaptor by incubating at 70 °C for 2 min and, subsequently, 37 °C for 5 min, followed by cooling to 4 °C.
6. Ligate 1 μl of a 5 μM 5′ RNA Adaptor solution by incubating with 10 U of T4 RNA Ligase 1 at 20 °C for 1 h.
7. Generate cDNA by incubation with 200 U of Superscript III Reverse Transcriptase at 50 °C for 1 h, followed by enzyme inactivation at 70 °C for 15 min and then cooling to 4 °C.
8. PCR-amplify the cDNA by incubation with 1 U of Phusion High-Fidelity DNA Polymerase, 25 pmol of Forward Primer and 25 pmol of Reverse PCR Primer, initially denaturing at 98 °C for 30 s, followed first by 15 PCR cycles of incubations at 98 °C for 10 s → 60 °C for 30 s → 72 °C for 30 s and, subsequently, by a final incubation at 72 °C for 10 min.
9. Purify PCR products from a nondenaturing 8% polyacrylamide gel. cDNA purification is essentially identical to the RNA purification as described in Subheading 3.4 except that TE Buffer and Phenol (pH 8.0) are used in place of water and TRIzol Reagent.
10. Evaluate the quality and quantity of cDNA using Agilent Bioanalyzer.
11. Sequence the cDNAs (6 pM in 1.2 ml) using a HiSeq 2500 Sequencing System with rapid run mode and the Sequencing Primer (0.5 μM in 1.2 ml). The total number of reads can be expected to be ~200 million.
12. Computationally analyze the sequencing data as shown in Fig. 3.

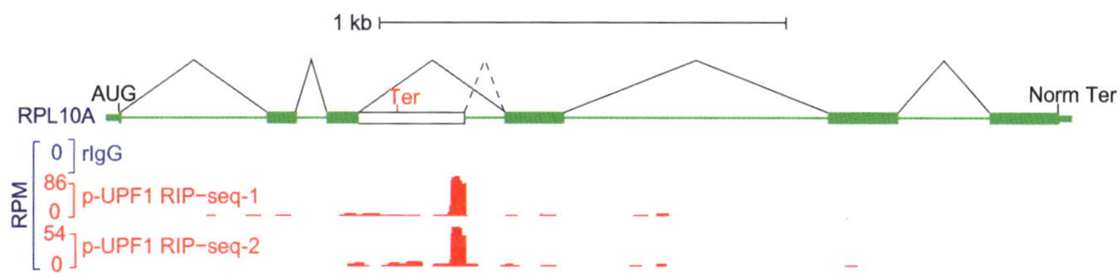

Fig. 3 Results from p-UPF1 RIP–seq demonstrate that the co-IP of p-UPF1 provides a useful identifier of cellular NMD targets. Example of p-UPF1 association with the splicing isoform of ribosomal protein L10a (RPL10A) mRNA that is a bona fide NMD target. Unspliced RPL10A pre-mRNA is diagramed, where sequences that constitute one or more mRNA isoforms are shown as *green* and *white boxes* separated by *green lines* that represent introns. AUG and Norm Ter denote the mRNA start codon and normal termination codon, respectively. The alternative splicing noted with a *dotted black line* generates a premature termination codon (Ter) in the white exon that triggers NMD [27, 28]. Two independent p-UPF1 RIP–seq experiments, labeled p-UPF1 RIP–seq-1 and p-UPF1 RIP–seq-2, demonstrate that p-UPF1 binding is enriched ~120 nt downstream of the Ter. This is evident by comparing the number of sequence reads per million mapped reads (RPM) for p-UPF1samples (*red*) compared to the number of sequence reads per million mapped reads for rabbit IgG samples (*blue*), the latter of which serve as negative controls by providing a background number of reads

4 Notes

1. Okadaic acid is a potent and specific inhibitor of type 1 and type 2A protein phosphatase [23]. Optimal conditions for okadaic acid use are specific to cell type and should be carefully established. The 100 μM stock aliquot can be stored at −20 °C.
2. These buffers should be freshly prepared. 1× Halt Protease and Phosphatase Inhibitor Cocktail (100×) (Thermo Fisher Scientific) can be used in place of 1× of a 7× cOmplete mini EDTA-free Protease Inhibitor cocktail tablet and 1× of a 10× PhosSTOP Phosphatase Inhibitor cocktail tablet.
3. Serine 1116 in the 1118-amino acid isoform of human UPF1 is equivalent to serine 1127 in the 1129-amino acid isoform of human UPF1. Both isoforms function in NMD and can vary in relative abundance depending on cell type [24].
4. A solution of 2-mercaptoethanol is readily oxidized and should be generated just prior to use.
5. The random four Ns in 5′ and 3′ adaptors help mitigate the bias of RNA ligases [25, 26].
6. Cell pellets can be stored at −80 °C for at least 6 months.
7. For a cell pellet of 100 mg, use ~1 ml of Hypotonic Gentle Lysis Buffer. This ratio of cells to the Lysis Buffer can vary dependent on cell type.
8. Use ~10% of cleared cell lysate for the total-RNA sample.
9. For washing, add 1 ml of buffer and rotate end-over-end ten times by hand until the beads are completely resuspended. Samples should be incubated on ice when using a Magnetic Tube Rack, waiting until all of the beads are captured on the side walls of the tubes before removing the buffer using an aspirator.
10. Use ~1 μl of Dynabeads Protein A/mg of cell pellet.
11. As an important negative control, RNAs captured by rabbit IgG-Dynabeads Protein A should be barely detectable (Fig. 3) relative to RNAs captured by rabbit IgG-Protein A Agarose.
12. Although we recommend using 5–10 μg of antibody for a lysate that contains 1–3 mg of total HEK293T-cell protein, this amount should be carefully tested for different cell types.
13. Conditions for the limited RNase I digestion should be optimized so that the bulk of RNA is <100 nt after digestion.
14. To optimize electrophoretic conditions, monitor in-gel RNA migration using the DynaMarker Prestain Marker for Small RNA.

15. This step is optional because the in-gel positions of RNAs migrating at 25–40 nt can be assessed using the DynaMarker Prestain Marker for Small RNA.
16. This step can be repeated, depending on the efficiency of sample elution.

Acknowledgments

We thank Keita Miyoshi for comments on the manuscript. Work on NMD in the Maquat lab is supported by NIH R01 GM059614 to L.E.M. T.K. was partially supported by a postdoctoral fellowship from FRAXA Research Foundation.

References

1. Lewis BP, Green RE, Brenner SE (2003) Evidence for the widespread coupling of alternative splicing and nonsense-mediated mRNA decay in humans. Proc Natl Acad Sci U S A 100:189–192. https://doi.org/10.1073/pnas.0136770100
2. Pan Q, Saltzman AL, Kim YK et al (2006) Quantitative microarray profiling provides evidence against widespread coupling of alternative splicing with nonsense-mediated mRNA decay to control gene expression. Genes Dev 20:153–158. https://doi.org/10.1101/gad.1382806
3. He F, Li X, Spatrick P et al (2003) Genome-wide analysis of mRNAs regulated by the nonsense-mediated and 5′ to 3′ mRNA decay pathways in yeast. Mol Cell 12:1439–1452. https://doi.org/10.1016/S1097-2765(03)00446-5
4. Kurosaki T, Maquat LE (2016) Nonsense-mediated mRNA decay in humans at a glance. J Cell Sci:461–467. https://doi.org/10.1242/jcs.181008
5. Mendell JT, Sharifi NA, Meyers JL et al (2004) Nonsense surveillance regulates expression of diverse classes of mammalian transcripts and mutes genomic noise. Nat Genet 36:1073–1078. https://doi.org/10.1038/ng1429
6. Gregersen LH, Schueler M, Munschauer M et al (2014) MOV10 is a 5′ to 3′ RNA helicase contributing to UPF1 mRNA target degradation by translocation along 3′ UTRs. Mol Cell 22:573–585. https://doi.org/10.1016/j.molcel.2014.03.017
7. Hogg JR, Goff SP (2010) Upf1 senses 3′UTR length to potentiate mRNA decay. Cell 143:379–389. https://doi.org/10.1016/j.cell.2010.10.005
8. Hurt JA, Robertson AD, Burge CB (2013) Global analyses of UPF1 binding and function reveal expanded scope of nonsense-mediated mRNA decay. Genome Res 23:1636–1650. https://doi.org/10.1101/gr.157354.113
9. Kurosaki T, Maquat LE (2013) Rules that govern UPF1 binding to mRNA 3′ UTRs. Proc Natl Acad Sci U S A 110:3357–3362. https://doi.org/10.1073/pnas.1219908110
10. Lee SR, Pratt GA, Martinez FJ et al (2015) Target discrimination in nonsense-mediated mRNA decay requires Upf1 ATPase activity. Mol Cell 59:413–425. https://doi.org/10.1016/j.molcel.2015.06.036
11. Zünd D, Gruber AR, Zavolan M et al (2013) Translation-dependent displacement of UPF1 from coding sequences causes its enrichment in 3′ UTRs. Nat Struct Mol Biol 20:936–943. https://doi.org/10.1038/nsmb.2635
12. Isken O, Kim YK, Hosoda N et al (2008) Upf1 phosphorylation triggers translational repression during nonsense-mediated mRNA decay. Cell 133:314–327. https://doi.org/10.1016/j.cell.2008.02.030
13. Cho H, Kim KM, Kim YK (2009) Human proline-rich nuclear receptor coregulatory protein 2 mediates an interaction between mRNA surveillance machinery and decapping complex. Mol Cell 33:75–86. https://doi.org/10.1016/j.molcel.2008.11.022
14. Kashima I, Yamashita A, Izumi N et al (2006) Binding of a novel SMG-1-Upf1-eRF1-eRF3 complex (SURF) to the exon junction complex triggers Upf1 phosphorylation and nonsense-mediated mRNA decay. Genes Dev

20:355–367. https://doi.org/10.1101/gad.1389006

15. Kurosaki T, Li W, Hoque M et al (2014) A post-translational regulatory switch on UPF1 controls targeted mRNA degradation. Genes Dev 28:1900–1916. https://doi.org/10.1101/gad.245506.114
16. Loh B, Jonas S, Izaurralde E (2013) The SMG5-SMG7 heterodimer directly recruits the CCR4-NOT deadenylase complex to mRNAs containing nonsense codons via interaction with POP2. Genes Dev 27:2125–2138. https://doi.org/10.1101/gad.226951.113
17. Okada-Katsuhata Y, Yamashita A, Kutsuzawa K et al (2012) N- and C-terminal Upf1 phosphorylations create binding platforms for SMG-6 and SMG-5:SMG-7 during NMD. Nucleic Acids Res 40:1251–1266. https://doi.org/10.1093/nar/gkr791
18. Anders KR, Grimson A, Anderson P (2003) SMG-5, required for *C. elegans* nonsense-mediated mRNA decay, associates with SMG-2 and protein phosphatase 2A. EMBO J 22:641–650. https://doi.org/10.1093/emboj/cdg056
19. Chiu S, Serin G, Ohara O et al (2003) Characterization of human Smg5/7a: a protein with similarities to *Caenorhabditis elegans* SMG5 and SMG7 that functions in the dephosphorylation of Upf1. RNA 9:77–87. https://doi.org/10.1261/rna.2137903
20. Ohnishi T, Yamashita A, Kashima I et al (2003) Phosphorylation of hUPF1 induces formation of mRNA surveillance complexes containing hSMG-5 and hSMG-7. Mol Cell 12:1187–1200. https://doi.org/10.1016/S1097-2765(03)00443-X
21. Durand S, Franks TM, Lykke-Andersen J (2016) Hyperphosphorylation amplifies UPF1 activity to resolve stalls in nonsense-mediated mRNA decay. Nat Commun 7:12434. https://doi.org/10.1038/nsmb.2575
22. Yamashita A, Ohnishi T, Kashima I et al (2001) Human SMG-1, a novel phosphatidylinositol 3-kinase-related protein kinase, associates with components of the mRNA surveillance complex and is involved in the regulation of nonsense-mediated mRNA decay. Genes Dev 15:2215–2228. https://doi.org/10.1101/gad.913001
23. Cohen P, Holmes CFB, Tsukitani Y (1990) Okadaic acid: a new probe for the study of cellular regulation. Trends Biochem Sci 15:98–102. https://doi.org/10.1016/0968-0004(90)90192-E
24. Nicholson P, Josi C, Kurosawa H et al (2014) A novel phosphorylation-independent interaction between SMG6 and UPF1 is essential for human NMD. Nucleic Acids Res 42:9217–9235. https://doi.org/10.1093/nar/gku645
25. Jayaprakash AD, Jabado O, Brown BD et al (2011) Identification and remediation of biases in the activity of RNA ligases in small-RNA deep sequencing. Nucleic Acids Res 39:e141. https://doi.org/10.1093/nar/gkr693
26. Zhuang F, Fuchs RT, Sun Z, Zheng Y, Robb GB (2012) Structural bias in T4 RNA ligase-mediated 3′-adapter ligation. Nucleic Acids Res 40:e54. https://doi.org/10.1093/nar/gkr1263
27. Mitrovich QM, Anderson P (2000) Unproductively spliced ribosomal protein mRNAs are natural targets of mRNA surveillance in *C. elegans.* Genes Dev 14:2173–2184. https://doi.org/10.1101/gad.819900
28. Takei S, Togo-Ohno M, Suzuki Y et al (2016) Evolutionarily conserved autoregulation of alternative pre-mRNA splicing by ribosomal protein L10a. Nucleic Acids Res 44:5585–5596. https://doi.org/10.1093/nar/gkw152

Chapter 14

Generation of Cell Lines Stably Expressing a Fluorescent Reporter of Nonsense-Mediated mRNA Decay Activity

Nadezhda M. Markina, Anton P. Pereverzev, Dmitry B. Staroverov, Konstantin A. Lukyanov, and Nadya G. Gurskaya

Abstract

Nonsense-mediated mRNA decay (NMD) is a mechanism of mRNA surveillance ubiquitous among eukaryotes. Importantly, NMD not only removes aberrant transcripts with premature stop codons, but also regulates expression of many normal genes. A recently introduced dual-color fluorescent protein-based reporter enables analysis of NMD activity in live cells. In this chapter we describe the method to generate stable transgenic cell lines expressing the splicing-dependent NMD reporter using consecutive steps of lentivirus transduction and Tol2 transposition.

Key words Nonsense-mediated mRNA decay, Splicing, Fluorescent proteins, Genetically encoded sensor, Fluorescence-activated cell sorting, Mammalian cell lines, Tol2 transposition

1 Introduction

All eukaryotic cells contain a complex pathway called nonsense-mediated mRNA decay (NMD), which recognizes and degrades mRNA species with a premature termination codon (PTC) [1]. The main sources of such aberrant transcripts with truncated coding regions are gene mutations, alternative splicing, and DNA rearrangement in immune cells.

PTC recognition is based on positioning of the terminating ribosome in improper contexts of two main types: (1) a long 3′ untranslated region (3′UTR) and (2) an intron more than 50 nucleotides downstream of the terminal codon [2]. The long 3′UTR-based recognition mechanism functions in all organisms studied, whereas splicing-dependent NMD is characteristic for vertebrates and plants [3, 4]. Such unusual convergence of very distant taxa implies that both NMD pathways originated at the earliest stages of evolution of eukaryotes [3].

Shireen R. Lamandé (ed.), *mRNA Decay: Methods and Protocols*, Methods in Molecular Biology, vol. 1720, https://doi.org/10.1007/978-1-4939-7540-2_14, © Springer Science+Business Media, LLC 2018

For decades, NMD was thought to be a housekeeping mechanism of mRNA quality control that prevents accumulation of potentially toxic C-terminally truncated proteins. Recent data, however, strongly suggest additional functions of NMD in global regulation of gene expression during key biological processes such as cell differentiation, embryonic development, and stress response [5]. This new understanding of biological NMD significance is based on two groups of facts. First, it was demonstrated that many normal wild type transcripts are degraded by NMD machinery because their terminal codons are recognized as PTC (e.g., due to an intron in the 3′UTR) [2, 5]. Second, NMD activity was found to undergo specific regulation, for example, by some microRNAs [6, 7], Ca^{2+} [8], or expression level of NMD factors [9, 10]. Thus, modulation of NMD activity can result in up- or down-regulation of hundreds of mRNA species [11–13] providing a way of orchestrated regulation of gene expression.

The proposed regulatory function of NMD calls for new methods to measure detailed spatio-temporal activity patterns of this cascade. Indeed, the classical techniques for evaluating mRNA abundance and stability, namely northern blot analysis and quantitative real-time PCR, work with large populations of cells and provide a very restricted resolution in space and time. This problem can be solved by bioluminescence and fluorescence reporters of NMD activity based on NMD-dependent expression of a luciferase or a fluorescent protein [8, 14, 15]. It enables evaluation of live cells down to single cell sensitivity, although precautions should be taken when interpreting the data, as the measurements are performed at the protein not mRNA level.

Recently, we designed a dual-color reporter of NMD activity based on ratiometric signal from two fluorescent proteins [16]. This reporter consists of two vectors. The first vector (pNMD+) expresses NMD-targeted (due to the presence of an intron in its 3′UTR) mRNA for the green fluorescent protein TagGFP2, as well as NMD-insensitive mRNA for the red fluorescent protein Katushka (*see* Fig. 1). The second vector (pNMD−) is identical to the first one except the distance between the TagGFP2 stop codon and the exon junction site has been shortened to 35 nucleotides, which makes this transcript insensitive to NMD. Red-to-green ratio in pNMD+-expressing cells is proportional to NMD activity, while control cells expressing pNMD− provide a reference green-to-red ratio characteristic for particular detection parameters (intensity and wavelengths of excitation beams, detection wavelengths and detector sensitivity, or exposure time).

As this NMD reporter gives estimates of NMD activity in single live cells, it potentially can be used in a variety of models to visualize changes in NMD activity in space and time using fluorescence microscopy or flow cytometry [16–18]. However, to date only

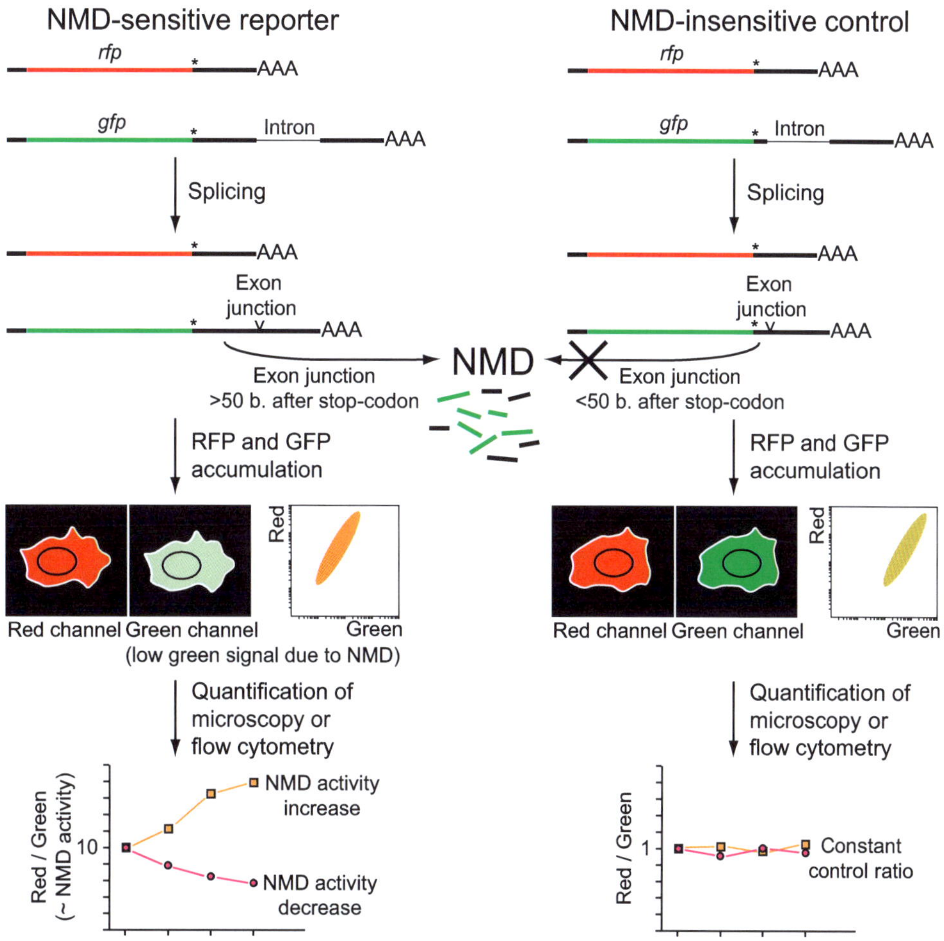

Fig. 1 Schematic outline of the NMD analysis method using dual-color fluorescence reporter. The splicing-dependent NMD pathway is evaluated by GFP-encoded transcript with an intron-containing 3′UTR (*left*). Control transcript contains the intron 35 nucleotides downstream of the stop codon and thus does not undergo NMD (*right*). RFP-encoding transcript acts as a reference for expression level. RFP and GFP fluorescence is detected by microscopy or flow cytometry, enabling quantification of NMD activity. For cells stably expressing the reporter, NMD activity can be followed in space and time during biological processes or treatments (*sketchy graphs at the bottom*)

transient expression of the NMD reporter has been described, which is inappropriate for long-term experiments.

In this chapter, we describe procedures to generate mammalian cell lines stably expressing the NMD reporter.

2 Materials

2.1 Genetic Vectors

1. Packaging lentivirus plasmids pR8.91 and pMD.G.
2. pLVT-Katushka transfer vector.

3. pCMVTol2 (AddGene) [19].
4. pTol2-NMD+ and pTol2-NMD− for Tol2-based stable integration (available upon request from the authors).
5. pNMD+ and pNMD− for transient transfections (available upon request from the authors).
6. pTurboFP635-N, pTagGFP2-N (Evrogen).
7. Standard *E. coli* strains, equipment and reagents for plasmid preparation.

2.2 Mammalian Cell Culture

Use standard protocols, equipment and reagents for cultivating mammalian cell lines. Use the HEK293T cell line for production of vector particles. The strongly adherent HeLa Kyoto cell line, adherent and suspension mouse Lewis Lung Carcinoma cells (LLC), adherent mouse Colon Cancer (CT26) can be used for both transient and stable expression experiments. Grow HeLa Kyoto and CT26 cells at 37 °C and 5% CO_2 in DMEM (Dulbecco's modified Eagle's medium) with 10% FBS (fetal bovine serum), 4 mM L-glutamine, 10 U/mL penicillin, 10 μg/mL streptomycin. For LLC cells use a mixture of DMEM and Ham's F12 medium (1:1).

For cell transfection, use appropriate common procedures using a transfection reagent, e.g., FuGENE HD (Promega) or electroporation, e.g., with Nucleofector 2b Device, (Lonza).

2.3 Cell Analysis

1. A Fluorescence-Activated Cell Sorter (FACS) machine.
2. A fluorescence microscope equipped with filter sets for detection in green and red channels.
3. A real-time PCR (qPCR) thermocycler.
4. Chemical inhibitors of NMD: caffeine, wortmannin, and/or NMDI14 (Sigma-Aldrich).

3 Methods

3.1 General Consideration on Choosing the Method of Transgenesis

The most convenient and efficient way to introduce an exogenous gene into genome of mammalian cells is recombinant lentiviral vectors [20]. However, this way is not well-suited for our NMD reporter as it contains an intron as well as two independent mRNAs. The life cycle of lentiviruses includes an intermediate mRNA-like form in the nucleus. As a result, spliceable introns are removed from the viral RNA. Also, the presence of transcription termination sites precludes formation of a full-length viral RNA genome. To solve this problem, antisense orientation of the insert can potentially be used, although often at expense of low virus titer [21, 22]. In our work, we failed to generate stable cell lines with NMD reporter using lentiviral vectors, possibly because there were cryptic splice sites or transcription terminators in the antisense orientation as well.

An alternative way to make transgenes is based on transposons. For example, the Tol2 transposon derived from *Oryzias latipes* (medaka fish) is autonomously active [23]. DNA of Tol2 consists of the transposase gene flanked by inverted terminal repeats required for transposition. Application of the Tol2 transposon system to generate stable expression in vertebrates including mammals has been reported [19, 24–26]. The smallest active variant *miniTol2* containing the combination of deletions has been developed [27]. The Tol2-mediated system was considered to have several advantages: (1) it has proved to be rather tolerant to increasing cargo size, which allows more complex transgenes to be incorporated without reducing transposition efficiency; (2) *Tol2* can transpose efficiently in a different cell types (either differentiating somatic cells or germ cells); (3) *Tol2* can integrate almost everywhere in the host genome, showing no preferences with respect to position on a chromosome; in contrast to other transposons, no consensus DNA sequence has been observed in the 8 bp target duplications created after integration of *Tol2;* (4) *Tol2* integration does not produce DNA rearrangements around its sites of integration [28].

As Tol2-based transgenesis is compatible with intron-containing and bicistronic inserts, we have chosen this system for our work. We have tested two alternative protocols to generate cell lines stably expressed NMD reporter. The first one used transgenesis by Tol2 vector with a long bicistronic insert carrying both intron-containing GFP- and intronless RFP-encoding genes. The potential advantage of this protocol is that it is based on a single-step transgenesis and ensures one-to-one copy numbers for GFP- and RFP-encoding genes. However, we encountered a significant problem of low stability of the integrated inserts. Even after careful selection of target dual-color transfectants by antibiotic selection and FACS, further cell passages resulted in a rather frequent appearance of green-only cells. We concluded that direct repeats close to each other in the bicistronic insert led to high probability of recombination events and elimination of the RFP-encoding part.

The second protocol used two consecutive steps (Fig. 2): (1) lentiviral transgenesis with the intronless RFP-encoding gene, and (2) Tol2-based transgenesis with intron-containing GFP-encoding genes (Fig. 3). This protocol includes more steps but provides efficient and stable transgenesis; it also provides flexibility in choosing a desirable green-to-red signal ratio. We have successfully used the suggested procedure for CT26 and LLC cells.

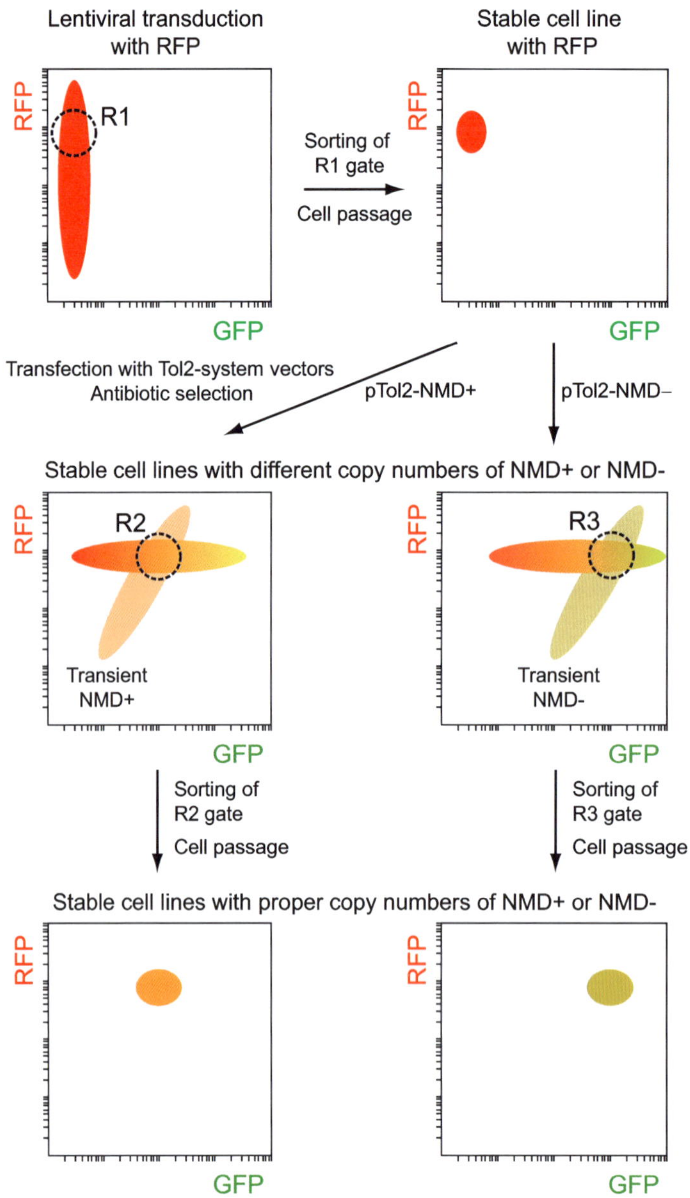

Fig. 2 Main steps to generate cell lines stably expressing the NMD reporter. Schematic flow cytometry bivariate plots in the green and red channels are shown to illustrate generation of RFP-expressing stable cell line by lentiviral transduction and FACS sorting (*upper panels*); subsequent integration of NMD-reporter using Tol2-based system (*middle panels*); and final FACS sorting of stably expressing cells (*bottom panels*) with proper green-to-red ratios chosen by analysis of cells transiently transfected with pNMD+ or pNMD− (regions R2 and R3, respectively)

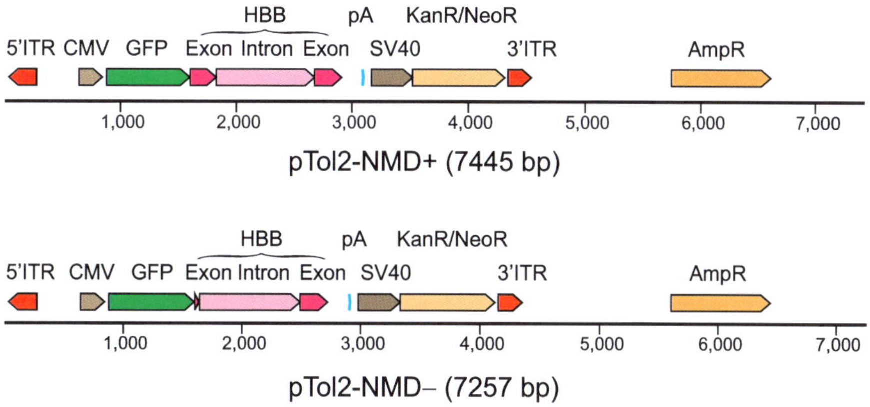

Fig. 3 Main elements of the reporter vectors pTol2-NMD+ and pTol2-pNMD−. Between inverted terminal repeats (5′ITR and 3′ITR) required for Tol2-based transposition, these vectors carry TagGFP2-encoding transcripts under control of CMV IE promoter. In their 3′UTR, these transcripts contain a fragment of human beta-globin (HBB) gene with a spliceable intron. In pTol2-NMD+, the intron is 230 bases downstream of the TagGFP2 stop-codon making this mRNA a classical target of splicing-dependent NMD. In pTol2-NMD−, the intron is only 35 bases downstream of the TagGFP2 stop-codon; thus this mRNA is not targeted to NMD

3.2 Generation of Katushka-Expressing Stable Cell Line by Lentiviral Transduction

In our method, Katushka provides a reference red signal for normalization. In fact, any RFP can be used for this purpose. Thus, if you already have a stable RFP-expressing cell line of interest (self-made or from a commercial source, e.g., from Anticancer Inc., www.anticancer.com), skip this section and go directly to the steps of Tol2-based integration of NMD reporter (Subheading 3.3).

3.2.1 Lentiviral Transfer Vector

The pLVT transfer vector was derived from the pRRLsin.cPPT.PGK.GFP.WPRE vector kindly provided by Prof. Didier Trono by replacing the PGK promoter with the intron-less version of the human elongation factor 1-α (EF1-α) promoter. The coding region of Katushka was excised from pTurboRFP-N and inserted into pLVT between BamHI and SalI restriction sites, replacing the GFP-encoding fragment, to produce the pLVT-Katushka transfer vector.

3.2.2 Production of Lentiviral Particles

Lentivirus particles can be produced by any convenient method. In our work we use the protocol according to [22, 29] with some modifications as described below.

1. Culture HEK293T cells in DMEM, supplemented with 10% FBS and antibiotics (50 U/mL penicillin G and 50 μg/mL streptomycin).
2. Seed out 1.5×10^6 HEK293T cells into a 60 mm culture dish 24 h before transfection.
3. Use manufacturer's protocol for transfecting HEK293T cells using FuGENE 6. We recommend using 4 μg and 1.2 μg of the

two packaging plasmids, pR8.91 and pMD.G, respectively, and 5 μg of transfer vector plasmid (pLVT-Katushka) for this transfection.

4. Replace the medium with 4 mL of fresh DMEM/10% FCS after 17 h.
5. After 24 h of incubation collect the medium with lentiviral particles and filter it through 0.45-μm filter.
6. Freeze the virus sample in 100–200 μL aliquots at −80 °C (without addition of DMSO). Use it for generating further stable cell lines of interest (*see* Subheading 3.2.3, **step 2**).
7. To determine the lentivirus titer, thaw one aliquot and prepare 2–3 serial tenfold dilutions of the virus in DMEM. Add each virus sample to 1×10^5 HeLa or HEK293T cells grown in 35-mm plastic dishes for 24 h in advance. Forty eight hours after transduction, count Katushka-positive (red fluorescent) cells by flow cytometry (*see* Subheading 3.2.4). To estimate the titer, use cell sample with 1–30% infection rate according to a standard protocol [30].

3.2.3 Lentiviral Transduction for Stable Expression of Katushka

The lentivirus sample should be used to generate cell line(s) of interest stably expressing Katushka. The protocol for CT26 and LLC cells given below should be adapted for a particular cell line (according to its size, growth rate, medium requirements, etc.).

1. Seed about $1–3 \times 10^5$ CT26 or $3–5 \times 10^4$ LLC cells into 35-mm plastic dishes the day before transduction.
2. For transduction, incubate the cells with virus-containing medium (*see* Subheading 3.2.2, **step 6**) using about 30 virus particles per cell for 24 h, then replace the medium.
3. Propagate the cells for 4–6 passages to reach a constant level of expression.

3.2.4 Analysis and Sorting of Cells with Stable Lentiviral Expression of Katushka by FACS (See Note 1)

1. To prepare cells for flow cytometry analysis, wash the cells twice with 1 mL of versene solution and trypsinize in 0.5–1 mL of trypsin/EDTA solution for 5 min at 37 °C. Centrifuge the required number of cells (10^6–10^7 cells per sample) at $300 \times g$ for 5 min at room temperature. Remove supernatant and carefully resuspend the cell pellet in 0.5–1 mL of PBS/0.5% BSA.
2. To remove cell aggregates, pass the sample through a 70-μm cell strainer immediately before sorting.
3. Analyse the cells by FACS. Apply commonly used flow cytometry gates in the forward/side scattering plots to select single live cells only.
4. Analyse cell fluorescence in green and red channels (e.g., 488 nm excitation and 510–530 nm detection for the green

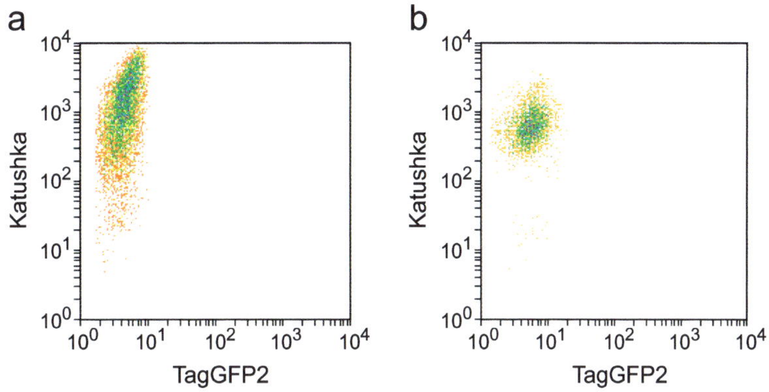

Fig. 4 Generation of a cell line stably expressing Katushka by lentiviral transduction. Shown are results of flow cytometry of CT26 cells after transduction with pLVT-Katushka (**a**) and further FACS sorting (**b**)

signal, and 561 nm excitation and 660–680 nm detection for the red signal; *see* Subheading 3.3.2 and **Note 1**). In a green-red bivariate plot (green versus red signal), the cells should give a vertical oval-shaped cloud (*see* Fig. 2, top left panel and Fig. 4a).

5. Sort a sufficiently bright (but not the brightest—*see* Fig. 2, gate R1) red cells under sterile conditions into a tube with 0.5 mL DMEM. Collect 1–5 × 10^4 cells (*see* **Note 2**). Seed the sorted cells into 35-mm dish or in a well of a 6-, 12-, or 24-well plate and grow for 96 h in DMEM with 20% FBS. If you collected less than 3 × 10^3 cells, seed in a 96-well plate and grow for 7 days.
6. Propagate the cells for 2–3 passages.
7. Repeat flow cytometry analysis at the same settings as above to prove that cells form a compact cloud within the previously selected region with no RFP-negative or dim cells in the sample (*see* Fig. 2, top right panel and Fig. 4b).
8. Freeze a portion of the obtained cell line stably expressing Katushka with DMSO in liquid nitrogen for future use.

3.3 Tol2-Based Integration of NMD Reporter

3.3.1 Tol2 Transposon Vectors

NMD reporter plasmids were constructed on the base of pminiTol2 vector [19] as follows: (1) neomycin-resistance (kanR/neoR) cassette under SV40 promoter was added; (2) fragment encoding either TagGFP2-HBB+ (NMD-sensitive) or TagGFP2-HBB− (NMD-insensitive) under control of CMV promoter was inserted to generate the final pTol2-NMD+ and pTol2-NMD− plasmids, respectively (*see* Fig. 3 and **Note 3**). Prepare sufficient amounts (20–50 μg) of pTol2-NMD+ and pTol2-NMD− plasmids using standard bacterial transformation, growth and ampicillin selection, and plasmid isolation protocols (*see* **Note 4**).

3.3.2 Cell Transfection and Antibiotic Selection

1. In preliminary experiments, determine the concentration of G418 that kills your cells of interest (usually 400–1000 μg/mL) (*see* **Note 5**).
2. Perform two separate transfections of the Katushka-expressing cells obtained at Subheading 3.2.4, **step** 7 with a mixture of (1) pTol2-NMD+ and pCMVTol2, (2) pTol2-NMD− and pCMVTol2. Use a molar ratio of 1.5:1 for pTol2-NMD+(−)/pCMVTol2 for cotransfection (i.e., taking into account their lengths, 500 ng of pTol2-NMD+(−) and 250 ng of pCMVTol2); higher or lower ratios result in decreased percentage of cells with stable integration. Use any common method of transfection (*see* **Note 6**).
3. Twenty four hours after transfection, trypsinize the cells and plate about 1/10 of the volume of cells onto 3.5-cm plates with complete medium (DMEM) supplemented with 100 μg/mL G418 (*see* **Note 5** and 7).
4. After 8–10 days of selection evaluate cells for the formation of foci (antibiotic-resistant cell clones). Foci may require an additional week or more to develop depending on the cell line and transfection/selection efficiency.
5. Trypsinize the cells and transfer about 1/5 of cells into new flask with fresh medium containing G418. Seed the cells at no more than 25% confluency, otherwise the rate of cell division decreases and antibiotic selection will not be efficient.
6. Replace antibiotic-containing medium every 4 days.
7. Collect cells by trypsinization and seed to additional culture flasks to obtain approximately 1×10^7 cells for further FACS sorting (Subheading 3.3.4). Gradually increase G418 concentration when replating cells in fresh media.

3.3.3 Fluorescence Microscopy of Cells

Fluorescence microscopy is a useful method to check progress in stable cell line generation. During antibiotic selection, analyze cell samples every 3 days to make sure that the cells fluoresce in both red and green channels (*see* **Note 8**). Use any microscope equipped with standard filter sets for GFP and RFP (*see* Spectra Viewer at http://evrogen.com/spectra-viewer/viewer.shtml to check the appropriateness of filters) (*see* **Note 9**). During first 10–14 days of selection, a rapid decrease in green fluorescent signal is expected because of gradual elimination of transiently introduced plasmids. The nascent cells with stable integration possess lower fluorescence than transiently transfected cells.

3.3.4 Flow Cytometry Analysis and Sorting of Cells with Stable Tol2-Based Integration of NMD Reporter

1. Two days before cell sorting, prepare 4 control cell samples that are transiently transfected with the following vectors: (a) pTagGFP2-N, (b) pTurboFP635-N, (c) pNMD+, (d) pNMD−. Transfect cells using the same protocol as in Subheading 3.3.2, **step 2** and incubate cells for 2 days.

2. Prepare cell suspensions (four control samples and two target cell samples with stable integration of pTol2-NMD+ and pTol2-NMD−) for flow cytometry analysis (*see* Subheading 3.2.4, **steps 1** and **2**).
3. Use control samples (a) and (b) to set up appropriate detection parameters in the green and red channels of the FACS machine that ensure well-detectable signals without cross talk of the green cells into red channel and vice versa. Apply commonly used flow cytometry gates in the forward/side scattering plots to select single live cells only. Keep these settings unchanged for all further cytometry analyses and sorting.
4. Analyze control samples (c) and (d). It should give diagonal clouds in the bivariate plot (green versus red signal); pNMD+ cell diagonal distribution is shifted to the region of lower green signal compared to the pNMD− cells, proportionally to the NMD activity (*see* Fig. 2, middle panels and Fig. 5a, b). These control samples provide proper green-to-red signal ratios corresponding to equal copy numbers of TagGFP2 and Katushka-encoding genes in NMD-sensitive and NMD-insensitive constructs, respectively.
5. Analyse cell sample pTol2-NMD+, which should give a horizontal oval-shaped cloud in the bivariate plot (*see* Fig. 2, left middle panel and Fig. 5c). This distribution corresponds to the presence of different numbers of integrated TagGFP2-encoding genes into the genome of the cells obtained after antibiotic selection.
6. Draw a region at the virtual intersection of pNMD+ and pTol2-NMD+ cell clouds (R2 in Fig. 2) (*see* **Note 10**). Sort pTol2-NMD+ cells from this region under sterile conditions into a tube with 0.5 mL DMEM. Collect 1–5 × 10^4 cells (*see* **Note 2**). Analogously, analyze cell sample pTol2-NMD−, which should give a horizontal oval-shaped cloud in the bivariate plot somewhat right-shifted compared to the pTol2-NMD+ cells (*see* Fig. 2, right middle panel and Fig. 5d).
7. Draw a region at the virtual intersection of pNMD− and pTol2-NMD− cell clouds (*see* Fig. 2, gate R3).
8. Sort pTol2-NMD+ cells from this region under sterile conditions into a tube with 0.5 mL DMEM. Collect 1–5 × 10^4 cells (*see* **Note 2**).
9. Seed the sorted cells into 35-mm dishes or into wells of a 6-, 12-, or 24-well plate and grow for 96 h in DMEM with 20% FBS.
10. Propagate the cells for 2–3 passages.
11. Repeat flow cytometry analysis at the same settings as above to prove that cells form a compact cloud within the previously

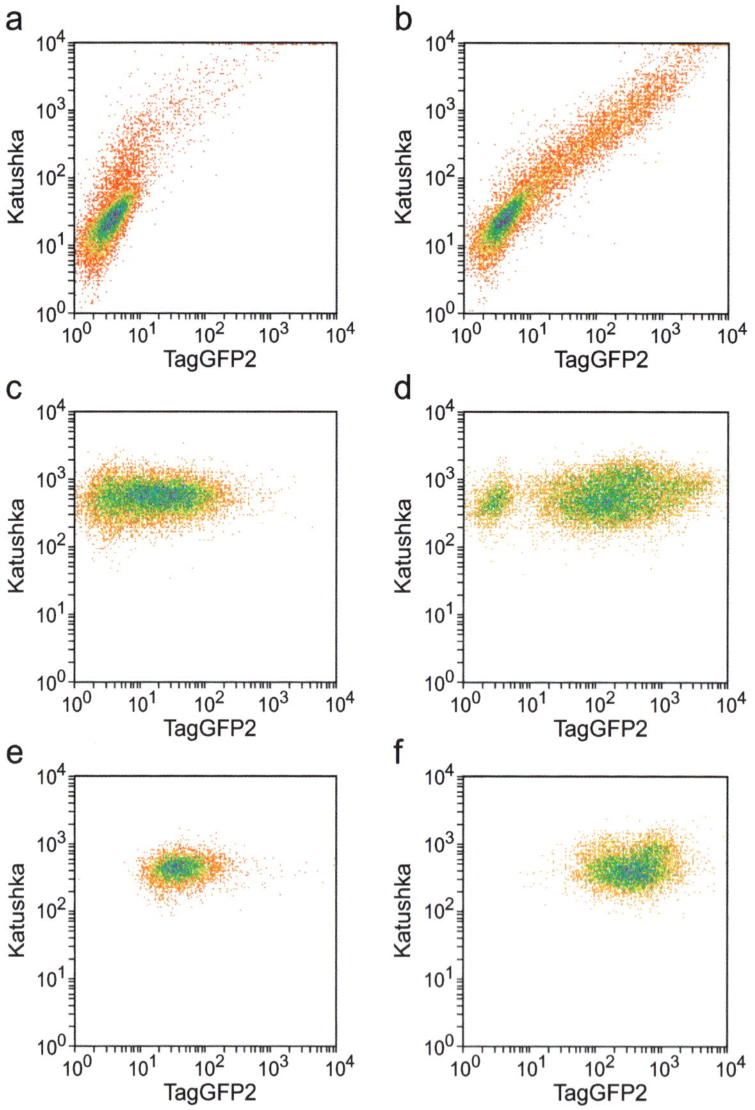

Fig. 5 Generation of cell lines stably expressing NMD reporter. Results of flow cytometry of CT26 cells are shown (the same settings for all panels). (**a**, **b**) Transient transfection with pNMD+ (**a**) or pNMD− (**b**) for estimation of green-to-red signal ratios corresponding to about 1:1 copy numbers of TagGFP2 and Katushka-encoding genes. (**c**, **d**) Tol2-based transposition of pTol2-NMD+ (**c**) or pTol2-NMD− (**d**) into Katushka-expressing stable cell line. (**e**, **f**) Katushka/pTol2-NMD+ (**e**) and Katushka/pTol2-NMD−(**f**) stable lines after FACS sorting (according to the chosen *green-to-red* ratios) and propagation

selected region with no RFP- or GFP-negative or very dim cells in the sample (*see* Fig. 2, bottom panels and Fig. 5e, f).

12. If required, repeat sorting of target cell populations (**steps 7–13**).
13. Freeze the obtained cell lines stably expressing Katushka and pTol2-NMD+ or pTol2-NMD− with DMSO in liquid nitrogen for future use.

3.4 Analysis of Stable Cell Lines

3.4.1 Inhibitory Analysis

To check whether the stably expressed NMD reporter correctly follows NMD activity, inhibitory analysis can be performed. Use treatments known to inhibit the NMD machinery such as:

1. Nonspecific (caffeine and wortmannin [31, 32]) or specific (NMDI14 [33]) NMD chemical inhibitors;
2. Inhibitory RNAs, e.g., shRNA against key NMD factor UPF1 [14, 15].

These treatments should result in significant suppression of NMD activity. Thus, an increase of green fluorescence in pTol2-NMD+-expressing cells is expected upon treatment. For each inhibition experiment use 4 samples of cells: NMD+ with inhibitor, NMD+ without inhibitor, NMD− with inhibitor, NMD− without inhibitor.

Treatment with Wortmannin

1. Seed out about 3–5 × 10^4 stably expressed cells (both Katushka/pTol2-NMD+ and Katushka/pTol2-NMD−) on 35-mm plastic dishes. Use two dishes for each cell culture.
2. Culture cells for 12–14 h in complete growth medium DMEM at 37 °C, 5% CO_2 in a humidified incubator.
3. Prepare 3 mL DMEM with 20 μM wortmannin by adding 30 μL of 2 mM wortmannin stock solution in DMSO. Also, prepare 3 mL DMEM with 30 μL DMSO (for vehicle controls).
4. Change cell medium to fresh prewarmed DMEM with wortmannin or DMSO in Katushka/pTol2-NMD+ and Katushka/pTol2-NMD− samples.
5. Culture cells for at least 24 h at 37 °C, 5% CO_2 in a humidified incubator.
6. Prepare 10^4–10^5 cells for flow cytometry analysis (*see* Subheading 3.2.4, **steps 1** and **2**).
7. Perform flow cytometry analysis (*see* Subheading 3.3.4, **steps 3** and **4**). Wortmannin treatment should result in a considerable increase of green signal (increase in green/red ratio) in Katushka/pTol2-NMD+ cells.
8. Estimate ratio of the green/red ratios in NMD+ and NMD− without wormannin (vehicle controls). Estimate the same value for wortmannin-treated samples. Difference between vehicle controls and wortmannin-treated samples provides a level of NMD inhibition.

Treatment with Caffeine

1. Seed out about 3–5 × 10^4 stably expressed cells (both Katushka/pTol2-NMD+ and Katushka/pTol2-NMD−) on 35-mm plastic dishes. Use two dishes for each cell culture.

2. Culture cells for 12–14 h in complete growth medium DMEM at 37 °C, 5% CO_2 in a humidified incubator.
3. Prepare 10 mM solution of caffeine powder (Sigma) in DMEM, sterilize it by filtering via 0.22 μm filter. Store this solution at +4 °C for up to 2 months.
4. Change cell medium in both Katushka/pTol2-NMD+ and Katushka/pTol2-NMD− samples to fresh prewarmed DMEM with 10 mM caffeine every 12 h. Replace cell medium to fresh DMEM without caffeine in control pair of samples Katushka/pTol2-NMD+ and Katushka/pTol2-NMD−.
5. Culture cells for at least 24 h.
6. Prepare 10^4–10^5 cells for flow cytometry analysis (*see* Subheading 3.2.4, **steps 1** and **2**).
7. Perform flow cytometry analysis (*see* Subheading 3.3.4, **steps 3** and **4**). To estimate the value of NMD inhibition, compare the ratio of green/red signal values for the treated and untreated NMD reporter samples.

3.4.2 DNA and mRNA Analysis

DNA and mRNA analysis by quantitative PCR can be used to quantify integration copy number and expression levels of Katushka- and TagGFP2-encoding genes according to standard protocols [18, 34, 35].

4 Notes

1. Fluorescence-activated cell sorting is a key approach for this protocol. It is a highly sophisticated technique, which requires complex and expensive equipment (cell sorter) and an experienced operator, who is able to conduct calibration and set up proper detection and gating parameters to sort high-purity target cell populations. Various FACS machines can be used according to common practice of sorting of live mammalian cells [36, 37]. Ideally, to avoid spectral cross talk, it should be equipped with blue (488 nm) and green-yellow (e.g., 561 nm) lasers focused at spatially separated spots for independent excitation of TagGFP2 and Katushka. If there is no laser in the 520–600 nm region in a particular FACS machine, single excitation line at 488 nm can be used for both TagGFP2 and Katushka [16, 18]. In this case, however, it is especially important to use control cells transfected with TagGFP2 or Katushka alone to select proper detection and compensation settings and avoid spectral cross talk between green and red signals. See Spectra Viewer at the Evrogen website (http://evrogen.com/spectra-viewer/viewer.shtml) to select optimal lasers and detection filters for TagGFP2 and Katushka (TubroFP635).

In our work we used FACSAria III (Becton Dickinson). TagGFP2 was excited with 488 nm laser and detected at 530/30 nm; Katushka was excited with 561 nm laser and detected at 670/14 nm.

2. You can use an option to sort single cells into 96- or 384-well plates. Single cell-derived clones require more time to grow but provide much more homogeneous fluorescence signals.
3. We use the CMV promoter to drive expression of the GFP-HBB cassette. This viral promoter is highly active in many cell lines (such as HeLa), but it is not ubiquitously expressed in primary cells [38]. To solve this problem, human PGK and EF-1α promoters can be used [39].
4. Nearly all commonly used *E. coli* strains (e.g., XL1-Blue) can be used for plasmid propagation. The quality of purified plasmid DNA may greatly affect transfection efficiency. In particular, treatment with endotoxin-removal buffer is very important to increase transfection efficiency for some cell lines (e.g., LLC).
5. We recommend determining optimal concentrations of antibiotic required to kill your host cell line before transfection with pTol2-NMD plasmid by treating the cells with several concentrations, ranging from 100 μg/mL to 1 mg/mL. Prepare a sterile 1600 μg/mL stock solution of G418 in PBS or DMEM and store it at 4 °C for several months. To find the minimum effective G418 concentration, a kill curve should be measured. After treatment, cell death occurs rapidly, allowing the selection of transfected cells with plasmids carrying the ***neo*** gene in a few passages.
6. In our work, we used FuGENE HD or FuGENE 6 reagents for transfecting CT26 cells, and electroporation (nucleofection) with Nucleofector 2b for difficult-to-transfect cells such as LLC. These procedures are described in detail in ref. 18.
7. Do not seed more than 1/10 of transfected cells (no more than 5×10^4 onto a plate with antibiotic). If the cells are plated with too high density, efficiency of antibiotic selection gradually decreases.
8. Monitoring cell fluorescence during selection of stable cell lines allows any red-negative green-positive cells that may appear to be removed. When a population of cells with fluorescence in only one channel is observed, use analysis and sorting of live cells by FACS. We do not recommend using this analysis before 14 days of G418 selection.
9. In our work we imaged cells with BZ9000 fluorescence microscope (Keyence) using 20× objective and standard filter sets for green (excitation at 450–490, emission at 510–560 nm) and red (excitation at 540–580, emission at 600–660 nm)

channels. We also used ZOE Fluorescent Cell Imager (Bio-Rad) with LED light sources.

10. Katushka/pTol2-NMD+ stably expressed cells with green-to-red ratio similar to that of transiently transfected pNMD+ cells carry close to one-to-one ratio of TagGFP2 and Katushka-encoding genes into their genome. Optionally, you can also select Katushka/pTol2-NMD+ cells with a higher level of green signal, which have more copies of integrated pTol2-NMD+ (*see* Fig. 2). Bright green fluorescence of these cells would be helpful in some biological models, e.g., analysis of tumors in vivo (where dim green signal is hard to detect reliably), or when increased NMD activity (and thus decreased green fluorescence) is expected during the experiment.

Acknowledgments

The authors are grateful to Tatiana Gorodnicheva (Evrogen) for providing pLVT-Katushka. This work was supported by the Russian Science Foundation grant 14-25-00129. The work was partially carried out using equipment provided by the Institute of Bioorganic Chemistry Core Facility (CKP IBCH, supported by Russian Ministry of Education and Science, grant RFMEFI62117X0018).

References

1. Chang YF, Imam JS, Wilkinson MF (2007) The nonsense-mediated decay RNA surveillance pathway. Annu Rev Biochem 76:51–74
2. Schweingruber C, Rufener SC, Zünd D, Yamashita A, Mühlemann O (2013) Nonsense-mediated mRNA decay—mechanisms of substrate mRNA recognition and degradation in mammalian cells. Biochim Biophys Acta 1829:612–623
3. Kertész S, Kerényi Z, Mérai Z, Bartos I, Pálfy T, Barta E, Silhavy D (2006) Both introns and long 3′-UTRs operate as cis-acting elements to trigger nonsense-mediated decay in plants. Nucleic Acids Res 34:6147–6157
4. Dai Y, Li W, An L (2016) NMD mechanism and the functions of Upf proteins in plant. Plant Cell Rep 35:5–15
5. Lykke-Andersen S, Jensen TH (2015) Nonsense-mediated mRNA decay: An intricate machinery that shapes transcriptomes. Nat Rev Mol Cell Biol 16:665–677
6. Bruno IG, Karam R, Huang L, Bhardwaj A, Lou CH, Shum EY, Song HW, Corbett MA, Gifford WD, Gecz J, Pfaff SL, Wilkinson MF (2011) Identification of a microRNA that activates gene expression by repressing nonsense-mediated RNA decay. Mol Cell 42:500–510
7. Jin Y, Zhang F, Ma Z, Ren Z (2016) MicroRNA 433 regulates nonsense-mediated mRNA decay by targeting SMG5 mRNA. BMC Mol Biol 17:17
8. Nickless A, Jackson E, Marasa J, Nugent P, Mercer RW, Piwnica-Worms D, You Z (2014) Intracellular calcium regulates nonsense-mediated mRNA decay. Nat Med 20:961–966
9. Chang L, Li C, Guo T, Wang H, Ma W, Yuan Y, Liu Q, Ye Q, Liu Z (2016) The human RNA surveillance factor UPF1 regulates tumorigenesis by targeting Smad7 in hepatocellular carcinoma. J Exp Clin Cancer Res 35:8
10. Sanchez G, Bondy-Chorney E, Laframboise J, Paris G, Didillon A, Jasmin BJ, Côté J (2016) A novel role for CARM1 in promoting nonsense-mediated mRNA decay: potential implications for spinal muscular atrophy. Nucleic Acids Res 44:2661–2676
11. Mendell JT, Sharifi NA, Meyers JL, Martinez-Murillo F, Dietz HC (2004) Nonsense

surveillance regulates expression of diverse classes of mammalian transcripts and mutes genomic noise. Nat Genet 36:1073–1078
12. Lykke-Andersen S, Chen Y, Ardal BR, Lilje B, Waage J, Sandelin A, Jensen TH (2014) Human nonsense-mediated RNA decay initiates widely by endonucleolysis and targets snoRNA host genes. Genes Dev 15:2498–2517
13. Schmidt SA, Foley PL, Jeong DH, Rymarquis LA, Doyle F, Tenenbaum SA, Belasco JG, Green PJ (2015) Identification of SMG6 cleavage sites and a preferred RNA cleavage motif by global analysis of endogenous NMD targets in human cells. Nucleic Acids Res 43:309–323
14. Boelz S, Neu-Yilik G, Gehring NH, Hentze MW, Kulozik AE (2006) A chemiluminescence-based reporter system to monitor nonsense-mediated mRNA decay. Biochem Biophys Res Commun 349:186–191
15. Paillusson A, Hirschi N, Vallan C, Azzalin CM, Muhlemann O (2005) A GFP based reporter system to monitor nonsense-mediated mRNA decay. Nucleic Acids Res 33:e54
16. Pereverzev AP, Gurskaya NG, Ermakova GV, Kudryavtseva EI, Markina NM, Kotlobay AA, Lukyanov SA, Zaraisky AG, Lukyanov KA (2015) Method for quantitative analysis of nonsense-mediated mRNA decay at the single cell level. Sci Rep 5:7729–7738
17. Pereverzev AP, Matlashov ME, Staroverov DB, Lukyanov KA, Gurskaya NG (2015) Differences of nonsense-mediated mRNA degradation activity in mammalian cell lines revealed by a fluorescence reporter. Bioorg Khim 41:587–591
18. Gurskaya NG, Pereverzev AP, Staroverov DB, Markina NM, Lukyanov KA (2016) Analysis of nonsense-mediated mRNA decay at the single-cell level using two fluorescent proteins. Methods Enzymol 572:291–314
19. Balciunas D, Wangensteen KJ, Wilber A, Bell J, Geurts A, Sivasubbu S, Wang X, Hackett PB, Largaespada DA, McIvor RS, Ekker SC (2006) Harnessing a high cargo-capacity transposon for genetic applications in vertebrates. PLoS Genet 2:e169
20. McCarron A, Donnelley M, McIntyre C, Parsons D (2016) Challenges of up-scaling lentivirus production and processing. J Biotechnol 240:23–30
21. Ismail SI, Kingsman SM, Kingsman AJ, Uden M (2000) Split-intron retroviral vectors: enhanced expression with improved safety. J Virol 74:2365–2371
22. Barde I, Salmon P, Trono D (2010) Production and titration of lentiviral vectors. Curr Protoc Neurosci 4(21):1–21
23. Kawakami K, Shima A (1999) Identification of the *Tol2* transposase of the medaka fish *Oryzias latipes* that catalyzes excision of a nonautonomous *Tol2* element in zebrafish *Danio rerio*. Gene 240:239–244
24. Sato Y, Kasai T, Nakagawa S, Tanabe K, Watanabe T, Kawakami K, Takahashi Y (2007) Stable integration and conditional expression of electroporated transgenes in chicken embryos. Dev Biol 305:616–624
25. Kawakami K, Noda T (2004) Transposition of the Tol2 element, an Ac-like element from the Japanese medaka fish Oryzias latipes, in mouse embryonic stem cells. Genetics 166:895–899
26. Koga A, Iida A, Kamiya M, Hayashi R, Hori H, Ishikawa Y, Tachibana A (2003) The medaka fish Tol2 transposable element can undergo excision in human and mouse cells. J Hum Genet 48:231–235
27. Urasaki A, Morvan G, Kawakami K (2006) Functional dissection of the Tol2 transposable element identified the minimal cis-sequence and a highly repetitive sequence in the subterminal region essential for transposition. Genetics 174:639–649
28. Kawakami K (2007) Tol2: a versatile gene transfer vector in vertebrates. Genome Biol 8 (Suppl 1):S7
29. Nakagawa T, Hoogenraad CC (2011) Lentiviral transgenesis. Methods Mol Biol 693:117–142
30. Liu YP, Berkhout B (2014) HIV-1-based lentiviral vectors. Methods Mol Biol 1087:273–284
31. Noensie EN, Dietz HC (2001) A strategy for disease gene identification through nonsense-mediated mRNA decay inhibition. Nat Biotechnol 19:434–439
32. Usuki F, Yamashita A, Higuchi I, Ohnishi T, Shiraishi T, Osame M, Ohno S (2004) Inhibition of nonsense-mediated mRNA decay rescues the phenotype in Ullrich's disease. Ann Neurol 55:740–744
33. Durand S, Cougot N, Mahuteau-Betzer F, Nguyen CH, Grierson DS, Bertrand E, Tazi J, Lejeune F (2007) Inhibition of nonsense-mediated mRNA decay (NMD) by a new chemical molecule reveals the dynamic of NMD factors in P-bodies. J Cell Biol 178:1145–1160
34. Jozefczuk J, Adjaye J (2011) Quantitative real-time PCR-based analysis of gene expression. Methods Enzymol 500:99–109
35. Nicholson P, Joncourt R, Mühlemann O (2012) Analysis of nonsense-mediated mRNA decay in mammalian cells. Curr Protoc Cell Biol 27:27.4.1–27.4.61

36. Pruitt SC, Mielnicki LM, Stewart CC (2004) Analysis of fluorescent protein expressing cells by flow cytometry. Methods Mol Biol 263:239–258
37. Zeyda M, Borth N, Kunert R, Katinger H (1999) Optimization of sorting conditions for the selection of stable, high-producing mammalian cell lines. Biotechnol Prog 15:953–957
38. Bovia F, Salmon P, Matthes T, Kvell K, Nguyen TH, Werner-Favre C, Barnet M, Nagy M, Leuba F, Arrighi JF, Piguet V, Trono D, Zubler RH (2003) Efficient transduction of primary human B lymphocytes and nondividing myeloma B cells with HIV-1-derived lentiviral vectors. Blood 101:1727–1733
39. Teschendorf C, Warrington KH Jr, Siemann DW, Muzyczka N (2002) Comparison of the EF-1 alpha and the CMV promoter for engineering stable tumor cell lines using recombinant adeno-associated virus. Anticancer Res 22:3325–3330

Chapter 15

Reactivation Assay to Identify Direct Targets of the Nonsense-Mediated mRNA Decay Pathway in *Drosophila*

Jonathan O. Nelson and Mark M. Metzstein

Abstract

Transcriptome analysis provides a snapshot of cellular gene expression and is used to determine how cells and organisms respond to genetic or environmental changes. Identifying the transcripts whose expression levels are regulated directly by the manipulation being examined from those whose expression changes as a secondary cause from the primary changes requires additional analyses. Here we present a technique used to distinguish direct targets of the nonsense-mediated mRNA decay (NMD) pathway in *Drosophila* from secondary gene expression effects caused by loss of this pathway. This technique uses pulsed reexpression of an essential NMD gene in *Drosophila* lacking this NMD factor, followed by analysis of the transcriptome over time. In this way, RNAs with a rapid reduction in expression upon reactivation of NMD activity, corresponding to primary NMD targets, can be identified. This technique could potentially be modified to identify direct targets of other mRNA decay mechanisms in *Drosophila* or other organisms.

Key words *Drosophila melanogaster*, Nonsense-mediated mRNA decay, NMD, *Upf2*, Reactivation, Transcriptome, RNA-seq

1 Introduction

Transcriptome analysis, either through microarray hybridization or by using next generation RNA sequencing, has become an essential tool for understanding how genetic and environmental manipulations affect cell function. When such experiments are initiated, the cell first responds by changing expression levels of transcripts that are directly under the control of the manipulated pathway. The altered expression of these primary targets typically induces other transcript level changes, causing many more genes to be misexpressed than those directly regulated by the pathway in question. While it is sometimes useful to know the complete repertoire of expression changes that occurs during a specific manipulation, the question most often being asked is what targets are directly regulated by the pathway being analyzed. Thus, it is important to have additional assays that can distinguish between direct and indirect

Shireen R. Lamandé (ed.), *mRNA Decay: Methods and Protocols*, Methods in Molecular Biology, vol. 1720, https://doi.org/10.1007/978-1-4939-7540-2_15, © Springer Science+Business Media, LLC 2018

targets of the pathways being studied. One such example is the use of CHIP-seq experiments to identify direct targets of transcription factors [1].

Nonsense-mediated mRNA decay (NMD) is a well characterized pathway that targets certain normal and defective transcripts to trigger their rapid degradation [2]. When NMD is inhibited, a large percentage of the genome is misexpressed [3–10]. As expected if a degradation pathway is inhibited, many of these RNAs show an increase in expression, however, an essentially equal number of RNAs show a decrease in expression, and such RNAs are unlikely to be direct NMD targets [3–10]. The widespread effects on expression levels are likely due to the increased expression of mRNAs directly targeted by the pathway, followed by upregulation and downregulation of other genes in response to the increased expression of the direct targets. Thus, to separate direct targets from secondary effects among the pool of genes with increased expression in cells or animals lacking NMD activity, additional assays have to be employed. One approach is to directly measure transcript decay rates, with the expectation that direct NMD targets should show increased stability upon pathway inactivation. The rate of mRNA decay can be measured by using transcription inhibitors to block de novo mRNA synthesis and measure the decrease in total mRNA abundance [11], or by pulse labeling mRNA with 5′-bromouridine and measuring the decreasing amount of these labeled mRNAs over time [10]. Both of these methods are well suited for cell culture studies, but less so for whole animals, due to the relatively slow rate of transcription inhibitor or 5′-bromouridine uptake that is achievable (primarily through feeding).

A very different approach that is much better suited for identifying direct NMD targets in whole organisms is a so-called “reactivation” assay, originally developed in yeast [12, 13], and subsequently modified for *Drosophila* [9]. In this assay, RNA levels are measured at steady state in organisms with a defect in a single gene required for NMD, and thus lack NMD-pathway activity. This step identifies both direct and indirect target gene increases. NMD is then “reactivated” by expression of a wild-type version of the defective gene, and RNA expression levels measured over time. The logic behind this approach is that direct targets are likely to rapidly decrease in expression, due to decay by NMD, while the levels of increased indirect targets will go down slowly in response to renormalization of the direct targets.

Here we present a detailed method to identify NMD targets genome-wide in *Drosophila* using reactivation of NMD activity. The major procedures for this method include: setting up of the appropriate genetic backgrounds for both experimental and control samples; RNA extraction; genome-wide expression profiling; and data analysis to distinguish direct from indirect NMD targets.

2 Materials

Solutions should be prepared at room temperature; no special handling of materials is required. Follow all waste disposal regulations when disposing of waste materials.

2.1 NMD Reactivation and Sample Collection

1. *Drosophila* genotypes: *y w Upf2*^{25G}*/FM7i, Act5C:GFP; UAS:Upf2* and *y w Upf2*^{25G}*/FM7i, Act5C:GFP;* + females. *FM7i, Act5C:GFP* / Y; *hsp70:GAL4* males [9, 14, 15].
2. Lay cage: An empty plastic *Drosophila* food bottle, with small holes poked into the top with a 20G or 18G needle (large enough for air to get in, but not so large that animals can enter and exit).
3. 30 mm petri dish lay plates: 2% agar, 2.5% sucrose, 0.5% Nipagin M in apple juice [16].
4. Yeast paste: Made by adding water (a few drops at a time) to 1 g of baker's yeast while mixing by hand with a spatula (*see* **Note 1**). Continue adding water and mixing until yeast becomes a smooth creamy texture. Make immediately prior to use.
5. Vials with standard *Drosophila* dextrose/cornmeal food and cotton stopper.
6. Small funnel (80 mm mouth to 15 mm opening).
7. 40 mm petri dishes.
8. Dissecting microscope equipped for monitoring GFP fluorescence.
9. 37 °C water bath.
10. Liquid N_2.

2.2 Analysis

1. Tophat software for RNA sequencing read alignment [17] and USeq package to identify differentially expressed genes [18] for RNA-seq quantitation, or *Drosophila* (V2) Gene Expression Microarray and GeneSpring GX (Agilent) software for microarray quantitation.
2. Analysis to identify reactivation targets requires basic data analysis software, such as Microsoft Excel or R.

3 Methods

All *Drosophila* are kept in vials containing standard dextrose/cornmeal *Drosophila* food [16]. All *Drosophila* are maintained at 25 °C unless otherwise noted.

3.1 Genetic Cross and Animal Collection

1. Collect *Drosophila* virgin females of the genotypes *y w Upf2*^{25G}*/FM7i, Act5C:GFP; UAS:Upf2* (the "Reactivation"

sample) and *y w Upf2*^{25G}*/FM7i, Act5C:GFP*; + (the "Control" sample). Place ~20 females of the Reactivation sample in a food vial with ~8 *FM7i, Act5C:GFP/Y*; *hsp70:GAL4* males. Do the same with the Control sample in a separate vial. Let this cross go for 1–2 days at 25 °C to allow mating to take place. Each cross should be performed in multiple replicates as controls for biological variation and to ensure enough animals are available for analysis (*see* **Note 2**).

2. After animals have mated for 1–2 days, spread a small amount of yeast paste on a lay plate using your finger. Prepare one plate for each condition.
3. Transfer each vial of *Drosophila* crosses to separate lay cages by placing a funnel in the mouth of the lay cage, with the open end of the cage facing up. Transfer flies from the vial to the cage using the funnel, tapping them to the bottom of the cage, and then quickly place a lay plate facing in on the mouth of the bottle (*see* **Note 3**). Seal the plate to the bottle with tape. Put the cages with the lay plate facing up into a 25 °C incubator.
4. Twenty four hours later, holding the cage with the lay plate at the top, tap the flies to the bottom of the cage and replace the lay plate with a new plate (spread with fresh yeast paste) (*see* **Note 4**). The original plate should visibly have laid eggs upon its surface. Place this plate in a covered Petri dish and store at 25 °C.
5. Twenty four hours later, use a fluorescent dissecting microscope to identify GFP-negative larvae (the genotype of these animals will be *Upf2*^{25G}/Y) from the lay plate with the eggs (*see* **Note 5**). Collect these larvae on a new lay plate with fresh yeast paste. Place the plate in a covered Petri dish and store at 25 °C.
6. Three days later, collect L3 animals from the plate of GFP-negative animals in a vial containing food. L3 animals can be distinguished by their large size and branched morphology of the anterior spiracles [19]. Place the animals directly in the food of the collection vial. Make sure to collect at least 40 animals to have ample animals for one replicate at each time point during the *Upf2* reactivation assay.

3.2 Upf2 Reactivation

1. Collect animals to be used for −0.5-h pre-heat-shock time point for each condition (5–10 larvae per sample) in a 1.5 mL microcentrifuge tube. Flash freeze collected animals in liquid N_2 and store at −80 °C until you are ready to isolate RNA.
2. Place vials with remaining animals in 37 ° C water bath for half an hour. Make sure that the food is completely submerged in the water (*see* **Note 6**). When time is complete, remove all animals from water bath and place back at 25 °C. Immediately collect animals for 0-h time point for each condition in a

1.5 mL microcentrifuge tube. Collected animals should be flash frozen in liquid N_2 and stored at −80 °C until RNA is isolated.

3. After 2 h of recovery from the heat shock, collect animals for the 2-h time point for each condition in a 1.5 mL microcentrifuge tube. Flash freeze the samples in liquid N_2 and store at −80 ° C until RNA is isolated.
4. After 4 h of recovery, collect animals for the 4-h time point for each condition in a 1.5 mL microcentrifuge tube (*see* **Note 7**). Again, collected animals should be flash frozen in liquid N_2 and stored at −80 °C until RNA is isolated.

3.3 RNA Isolation and Quantitation

1. RNA can be isolated from samples using any technique RNA isolation protocol the user prefers that is suitable for RNA sequencing, e.g., *see* doi:https://doi.org/10.2506/cgbtr-200610
2. Quantitation can be performed using RNA sequencing or microarray analysis.

3.4 Analysis

1. Measure expression levels for RNAs:
   (a) For RNA sequencing, align reads to the *Drosophila* genome using the Tophat program and identify differentially expressed genes using the USeq package.
   (b) For microarrays, use appropriate software depending on the source of the microarrays (we used Agilent *Drosophila* (V2) Gene Expression Microarray and GeneSpring GX software).
2. Filter out genes whose expression is too low to be reliably quantitated (e.g., fewer than ten reads/transcript for RNA-Seq data).
3. Filter out genes with expression changes greater than 20% from the −0.5 time point to the 4 h time point in the *y w* $Upf2^{25G}$/Y; *hsp70:GAL4/+* condition. This will remove those effects that the heat shock treatment alone has on expression, and are thus not specific to *Upf2* reactivation.
4. Filter out genes that differed in expression by greater than 50% between the Reactivation and Control conditions at the −0.5 time point. These expression differences are mostly the result of leaky GAL4 expression, causing a small amount of the *Upf2* rescue construct to be expressed.
5. Normalize the levels of each transcript over the time course to the −0.5 time point, log transform, and calculate decay constant for Reactivation and Control conditions by linear fitting to these log transformed expression ratios. If transcripts are undergoing decay, this value should be negative.

6. Calculate the difference between Reactivation and Control slopes for each transcript, a value called “relative slope”.
7. Identify genes with decreased expression of 1.8 fold or higher over the 4 h time course in the Reactivation condition.
8. Direct targets of the NMD pathway (“reactivation targets”) are defined as those transcripts whose relative slope is negative and with decreased expression of >1.8 fold over the 4 h time course in the Reactivation condition.

4 Notes

1. It is easiest to prepare the yeast paste in a 40 mm Petri dish. A larger volume of paste can also be made in a small beaker, and stored covered with Parafilm at 4 °C for 1–2 days.
2. When mating the Reactivation and Control female animals to the +/Y; *hsp70:GAL4* males, it is helpful to have multiple replicates of each mating to ensure enough larvae for future steps.
3. It may be easier to first place the lay plate in the bottle and tape it in place, and then quickly remove it when adding the flies to the bottle, followed by putting the lay plate back in place.
4. Lay plates can become overcrowded, which may make distinguishing GFP fluorescence and recovering individual animals difficult, as well as delay development to the L3 stage. After setting up lay cages, changing and collecting lay plates twice a day will reduce the density of animals on those plates making collections easier.
5. When collecting GFP-negative animals, it can be difficult under some microscopes to distinguish GFP-positive from GFP-negative animals at early stages of development because the larvae are so small. In this case, waiting a further 24 h to collect larvae will make it easier to distinguish GFP fluorescence.
6. Animals may crawl up the sides of the vial during the heat shock to avoid the heat. To insure they stay at the bottom of the vial, push the cotton vial stopper almost completely to the bottom of the vial, only leaving a small space between the stopper and the food. Make sure that all of the area between the bottom of the vial and the stopper are submerged in the water bath to ensure all animals are exposed to the heat.
7. The 2- and 4-h collections are the standard time points we have used. More frequent time points could be used if desired, but these extra time points would not necessarily provide greater resolution for slope analysis.

References

1. Boeva V (2016) Analysis of genomic sequence motifs for deciphering transcription factor binding and transcriptional regulation in eukaryotic cells. Front Genet 7:e1000916
2. Karousis ED, Nasif S (2016) Nonsense-mediated mRNA decay: novel mechanistic insights and biological impact. WIREs RNA 7:661–682
3. He F, Li X, Spatrick P et al (2003) Genome-wide analysis of mRNAs regulated by the nonsense-mediated and 5′ to 3′ mRNA decay pathways in yeast. Mol Cell 12:1439–1452
4. Lewis BP, Green RE, Brenner SE (2003) Evidence for the widespread coupling of alternative splicing and nonsense-mediated mRNA decay in humans. Proc Natl Acad Sci U S A 100:189–192
5. Rehwinkel J, Letunic I, Raes J et al (2005) Nonsense-mediated mRNA decay factors act in concert to regulate common mRNA targets. RNA 11:1530–1544
6. Ramani AK, Nelson AC, Kapranov P et al (2009) High resolution transcriptome maps for wild-type and nonsense-mediated decay-defective *Caenorhabditis elegans*. Genome Biol 10:1
7. Hansen KD, Lareau LF, Blanchette M et al (2009) Genome-wide identification of alternative splice forms down-regulated by nonsense-mediated mRNA decay in *Drosophila*. PLoS Genet 5:e1000525–e1000514
8. Barberan-Soler S, Lambert NJ, Zahler AM (2009) Global analysis of alternative splicing uncovers developmental regulation of nonsense-mediated decay in *C. elegans*. RNA 15:1652–1660
9. Chapin A, Hu H, Rynearson SG et al (2014) *In vivo* determination of direct targets of the nonsense-mediated decay pathway in *Drosophila*. G3 (Bethesda) 4:485–496
10. Tani H, Imamachi N, Salam KA et al (2014) Identification of hundreds of novel UPF1 target transcripts by direct determination of whole transcriptome stability. RNA Biol 9:1370–1379
11. Ayupe AC, Reis EM (2017) Evaluating the stability of mRNAs and noncoding RNAs. In: Enhancer RNAs, Methods and protocols. Humana Press, New York, pp 139–153
12. Maderazo AB, Belk JP, He F, Jacobson A (2003) Nonsense-containing mRNAs that accumulate in the absence of a functional nonsense-mediated mRNA decay pathway are destabilized rapidly upon its restitution. Mol Cell Biol 23:842–851
13. Johansson MJO, He F, Spatrick P et al (2007) Association of yeast Upf1p with direct substrates of the NMD pathway. Proc Natl Acad Sci U S A 104:20872–20877
14. Reichhart JM, Ferrandon D (1998) Green balancers. Drosoph Inf Serv 81:201–202
15. Metzstein MM, Krasnow MA (2006) Functions of the nonsense-mediated mRNA decay pathway in *Drosophila* development. PLoS Genet 2:e180
16. Sullivan W, Ashburner M, Hawley RS (2000) *Drosophila* protocols. Cold Spring Harbor Laboratory Press, New York
17. Trapnell C, Roberts A, Goff L et al (2012) Differential gene and transcript expression analysis of RNA-seq experiments with TopHat and Cufflinks. Nat Protoc 7:562–578
18. Nix DA, Courdy SJ, Boucher KM (2008) Empirical methods for controlling false positives and estimating confidence in ChIP-Seq peaks. BMC Bioinformatics 9:523
19. Ashburner M, Golic KG, Hawley KG (2005) *Drosophila*: a laboratory handbook. Cold Spring Harbor Laboratory, New York

Chapter 16

Studying Nonsense-Mediated mRNA Decay in Mammalian Cells Using a Multicolored Bioluminescence-Based Reporter System

Andrew Nickless and Zhongsheng You

Abstract

The nonsense-mediated mRNA decay (NMD) pathway degrades aberrant transcripts containing premature translation termination codons (PTCs) and also regulates the levels of many normal mRNAs containing NMD-inducing features. The activity of this pathway varies considerably in different cell types and can change in response to developmental and environmental cues. Modulating NMD activity represents a potential therapeutic avenue for certain genetic disorders and cancers. Simple reporter systems capable of faithfully assessing NMD activity in mammalian cells greatly facilitate both basic and translational research on NMD. Here we describe a simple and effective method for assaying NMD specifically and quickly in live mammalian cells using a multicolored bioluminescence-based reporter system. This reporter can be transiently or stably introduced into cultured cells as well as animals, and NMD activity can be accurately assessed by bioluminescence imaging, western blot, or RT-qPCR.

Key words Nonsense-mediated mRNA decay (NMD), NMD reporter, CBR, CBG, Bioluminescence imaging, qPCR, Western blotting

1 Introduction

Nonsense-mediated mRNA decay (NMD) ensures appropriate gene expression by eliminating aberrant mRNAs containing premature termination codons (PTCs) and by regulating the abundance of many normal transcripts that possess features recognizable to the NMD machinery (e.g., 3′ UTR introns, upstream open reading frames, long 3′ UTRs) [1–4]. NMD contributes to diverse physiological processes and manipulation of this pathway is a promising therapeutic strategy for treating genetic disorders and cancers [3, 5]. The NMD machinery in humans is comprised of 18 NMD factors identified thus far, which include the UPF proteins, originally identified in yeast, and the SMG proteins, originally identified in *C. elegans* [6]. To mechanistically dissect the NMD pathway, explore its physiological significance, and identify drugs capable of

Shireen R. Lamandé (ed.), *mRNA Decay: Methods and Protocols*, Methods in Molecular Biology, vol. 1720,
https://doi.org/10.1007/978-1-4939-7540-2_16,

modulating NMD activity, simple assays capable of accurately measuring NMD activity are imperative. Efforts in this direction have led to the development of the fluorescent EGFP-TCRβ(PTC) reporter and bioluminescent RLuc-β-globin(PTC) reporter, in which EGFP and Renilla luciferase are fused to a PTC-containing TCRβ minigene and a β-globin gene, respectively [7, 8]. Wild type EGFP-TCRβ(PTC) or firefly luciferase (FLuc) in a separate vector is used as an internal control for DNA delivery and gene expression [7, 8]. We have recently developed a highly effective, single-vector NMD reporter system by taking advantage of the unique features of click beetle luciferases [9].

Our NMD reporter consists of two separate but highly homologous transcription units inserted in tandem in a single vector (Fig. 1a). Each unit contains a luciferase inserted into a TCRβ

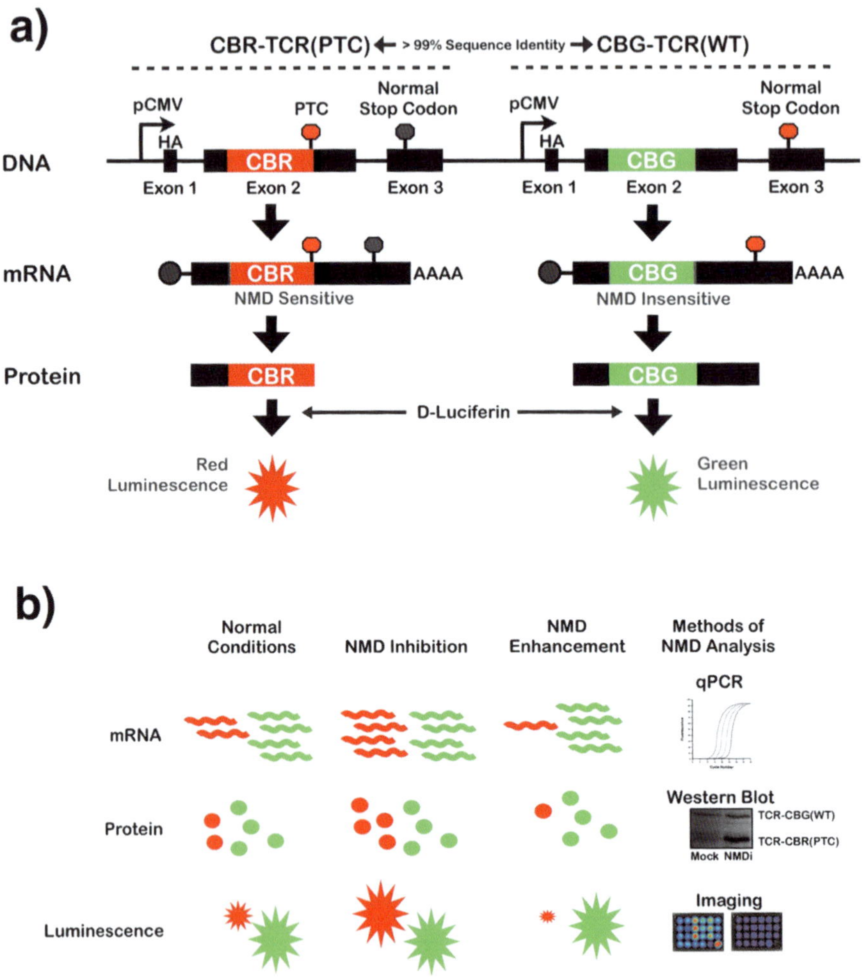

Fig. 1 A dual-color bioluminescent NMD reporter. (**a**) Schematic of the dual-color bioluminescent NMD reporter system. (**b**) Illustration of how the reporter responds to NMD modulation and the methods to detect NMD

minigene containing three exons and two introns. In the first transcription unit (CBR-TCR(PTC)), which encodes a PTC-containing nonsense mRNA, the click beetle red luciferase (CBR) was inserted into the second exon of TCRβ minigene with the natural stop codon of CBR serving as a PTC. PTCs in the well-characterized TCRβ minigene are particularly robust NMD substrates, although a small percentage of transcripts evade degradation [10]. In the second unit (CBG-TCR(WT)), which encodes a wild type transcript, the click beetle green luciferase CBG99 lacking its natural stop codon was inserted into the same position in the second exon of TCRβ minigene. Expression of both reporter genes is controlled by separate cytomegalovirus (CMV) promoters, splicing sites, and polyadenylation signals of identical sequences. Furthermore, both reporters share >99% sequence identity at the DNA, pre-mRNA, and mRNA level.

This versatile reporter enables NMD quantification via qPCR, western blotting, and bioluminescence imaging. The dual-reporter system provides an accurate, internally controlled metric for NMD quantification—the CBR–CBG ratio. Increases in this ratio indicate NMD repression while decreases indicate NMD enhancement (Fig. 1b). This ratio is most simply determined by bioluminescence imaging, as CBR and CBG signal intensity reliably reflect the steady state levels of the NMD target CBR-TCR(PTC) reporter mRNA and the non-NMD target CBG-TCR(WT) reporter mRNA. The use of the CBG-TCR(WT) internal control reporter gene minimizes nonspecific effects from variations in reporter DNA delivery, cell viability, or gene expression. Thus, changes in the CBR–CBG ratio reflect authentic differences in NMD efficiency of the reporter. In addition to minimizing indirect effects, the highly related CBR and CBG luciferases enable rapid measurement of NMD with the same D-luciferin substrate in a single reaction. A spectral deconvolution algorithm has been developed to unmix the bioluminescence signals of CBR and CBG, whose emission spectra partially overlap [11] (Fig. 2). This algorithm uses values from samples expressing only the CBR or CBG luciferase to mathematically remove the signal overlap in samples expressing both luciferases [11]. Because of the aforementioned design and the features of luciferase reactions, this reporter system is highly specific, sensitive, and robust, and is capable of measuring NMD activity in live cells and even small animals in a noninvasive manner. Furthermore, this convenient reporter system is well suited to high throughput screening to identify small molecule inhibitors or enhancers of NMD or to identify new NMD factors or regulators in combination with RNAi or CRISPR/Cas9 technologies.

In addition to bioluminescence imaging, NMD of the reporter can be measured using western blot or RT-qPCR (Fig. 1b). A hemagglutinin (HA) tag sequence was inserted into the first exon of both of the CBR-TCR(PTC) and CBG-TCR(WT) reporter

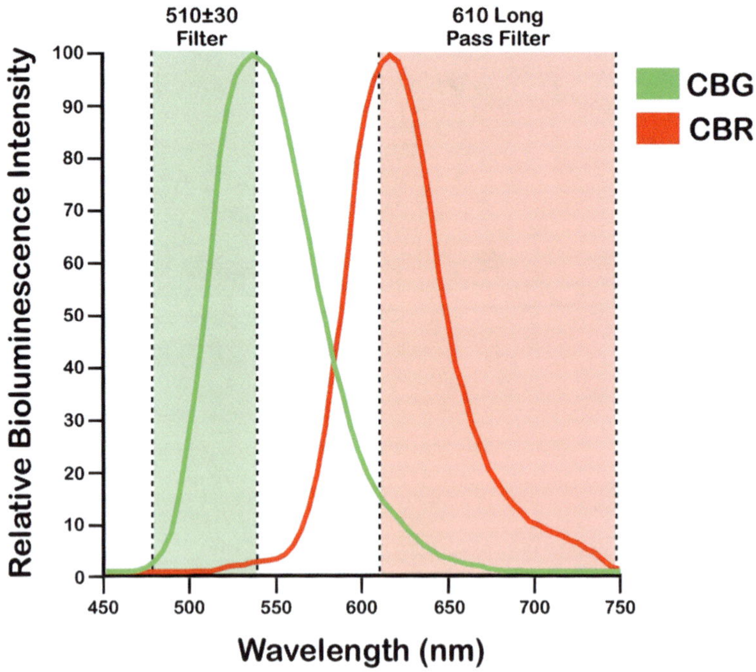

Fig. 2 Emission spectra of CBG and CBR luciferases

genes, which allows detection of the levels of the corresponding translated reporter protein products on a single western blot. HA immunoblotting also provides information about the sizes of reporter protein products, which reflects whether the reporter pre-mRNAs are properly spliced or whether translational read-through occurs on the CBR-TCR(PTC) reporter mRNA under specific conditions. Besides these indirect assays, NMD can be measured directly through RT-qPCR using primers targeting two regions in the reporter mRNAs that are distinct between CBR and CBG.

In this chapter, we describe a step-by-step procedure for introducing this reporter into cells and using it to accurately measure NMD in living cells by performing bioluminescence imaging, western blotting, and RT-qPCR. We also outline methods to process and analyze acquired data.

2 Materials

2.1 Cell Transfections and Selection

1. TransIT-LT1 Transfection Reagent (Mirus).
2. Opti-MEM (Invitrogen).
3. Plasmids:
   (a) pBS-(CBR-TCR(PTC)).

(b) pBS-(CBG-TCR(WT)).
(c) pBS-(CBR-TCR(PTC)-CBG-TCR(WT)).
(d) pMX-puromycin or similar puromycin-resistance-conferring plasmid.
(e) pAdenox-zsGreen-(CBR-TCR(PTC)-CBG-TCR(WT)).

4. 10 mg/mL puromycin dissolved in H_2O and stored at −20 °C.
5. TE buffer: 10 mM Tris–HCl pH 8.0, 0.1 mM EDTA.
6. Sterile cloning disks (Scienceware).

2.2 NMD Reporter Bioluminescence Imaging Assay

1. Microplate reader or CCD camera-base imaging system equipped with optical filters.
2. Black-walled cell culture dishes.
3. 30 mg/mL D-Luciferin (Goldbio) dissolved in H_2O and stored at −20 °C.
4. Phenol Red-free colorless medium appropriate for cell type.
5. Phosphate buffered saline (PBS) or TD buffer (138 mM NaCl, 5 mM KCl, 0.7 mM Na_2HPO_4, 25 mM Tris, pH 7.4).

2.3 NMD Reporter Western Blotting

1. HA tag antibody (Convance, MMS-101R).
2. SDS sample buffer: 50 mM Tris–HCl, pH 6.8, 10% glycerol, 2% SDS, 5% β-mercaptoethanol, bromophenol blue.
3. PDVF Membrane.
4. Sonic Dismembranator with sonicator probe (Fisher Scientific Sonic Dismembrator Model 100 probe sonicator or comparable model).

2.4 NMD Reporter RNA Analysis

1. RNA column based isolation kit.
2. cDNA synthesis kit.
3. SYBR Green qPCR reagent.
4. Optical plates.
5. Optical film or optical cap.
6. qPCR machine.
7. qPCR primers for CBG and CBR.

(a) CBG For: atgctctcgatccacgcgtg	CBG Rev: cgagagagtgaatgttagcg
(b) CBR For: tccatgctttcggctttcat	CBR Rev: cgagagtctggataatcgca
(c) GAPDH For: cctgttcgacagtcagccg	GAPDH Rev: cgaccaaatccgttgactcc
(d) HPRT For: tgacactggcaaaacaatgca	HPRT Rev: ggtccttttcaccagcaagct
(e) β-Actin For: gagcgcggctacagct	β-Actin Rev: tccttaatgtcacgcacgattt

3 Methods

3.1 Transfecting Human Cells with Reporter Plasmids

1. One day prior to transfection, seed human cells (e.g., U2OS) so that they will reach ~70% confluency the following day.
2. Transfect cells with NMD reporter plasmid (pBS-(CBR-TCR (PTC)), pBS-(CBG-TCR(WT)), or pBS-(CBR-TCR(PTC))-(CBG-TCR(WT))) using TransIT-LT1 reagent according to manufacturer's instructions.
3. Proceed to imaging and/or sample collection (Subheadings 3.4–3.6) within 48–72 h of transfection to maximize expression.

3.2 Generating NMD Reporter Adenoviruses and Target Cell Infection

1. Linearize 6 μg of pAdenox-zsGreen-(CBR-TCR(PTC)-CBG-TCR(WT)) plasmid with PacI digestion to expose inverted terminal repeats. This digestion will release a ~3 kb section of plasmid. Run 10% of reaction on a 1% agarose gel to ensure efficient digestion.
2. Isolate DNA by phenol–chloroform extraction and resuspend DNA in 10 μL TE buffer.
3. Transfect the linearized DNA into HEK293 cells growing in a 60 mm culture dish using TransIT-LT1 transfection reagent according to the manufacture's protocol. Cells should be 50–70% confluent at time of transfection (plated 18–24 h prior to transfection).
4. Check cells daily for evidence of the cytopathic effect, whereby cells become round and detach from the plate. When >50% of cells have detached from the plate, proceed to **step 5**; this often takes 2–3 weeks but generally occurs more quickly when amplifying virus (**step 9**). Do not change medium. Follow appropriate adenoviral safety protocols.
5. Transfer cells and medium to a sterile 15 mL conical tube.
6. Centrifuge at 1500 × *g* for 5 min (room temperature), remove medium, and resuspend cells in 500 μL of PBS or TD buffer.
7. Cell lysis: place the tube in a mixture of dry ice and 70% ethanol until the solution is solidified. Thaw in 37 °C water bath and vortex. Do not let solution reach 37 °C. Repeat three times.
8. Centrifuge lysate at 1500 × *g* for 5 min (room temperature) to pellet debris and transfer the virus-containing supernatant to a fresh microfuge tube.
9. Amplify virus: Remove medium from 10 cm dish of HEK293 cells (~70% confluent) and replace with 10 mL of medium containing 100 μL of the lysate from **step 7** and repeat **steps 4–8**. Store virus at −80 °C or proceed to **step 10**.

10. Remove medium from experimental cells and replace with medium containing reporter adenovirus at least 24 h prior to reporter imaging or sample collection (Subheadings 3.4–3.6) (*see* **Notes 1** and **2**).

3.3 Generating Stable Cell Lines Expressing the NMD Reporter

1. Cotransfect cells with the reporter plasmid (pBS-(CBR-TCR(PTC)), pBS-(CBG-TCR(WT)), or pBS-(CBR-TCR(PTC))-(CBG-TCR(WT))) (*see* Subheading 3.1) and an expression plasmid encoding a puromycin-resistance gene at a 9:1 ratio using the TransIT-LT1 transfection reagent.
2. Plate diluted cells and select single colonies with puromycin (1.5 μg/mL for U2OS cells) for 14 days.
3. Collect single colonies using sterile cloning disks soaked in trypsin solution according to the manufacturer's instructions. Expand culture.
4. Verify reporter expression with western blot, bioluminescence imaging, and RT-qPCR (*see* Subheadings 3.4–3.6). Use untransfected cells as a negative control for expression of the reporter genes.
5. Validate reporter function: Chemically (e.g., with caffeine, an inhibitor of the SMG1 kinase) and/or genetically (e.g., UPF1 or SMG1 knockdown) alter NMD activity and assess changes in the CBR–CBG ratio via western blot, bioluminescence imaging, and RT-qPCR. Inhibition of NMD by these means will increase the CBR–CBG ratio as the PTC-containing mRNA is stabilized (Fig. 1b). If the reporter is expressed and functions correctly, cells are ready for experimental manipulation and NMD analysis (Subheadings 3.4–3.6).

3.4 Bioluminescence Imaging and Data Analysis (See Notes 3 and 4)

1. Dilute D-luciferin (30 mg/mL stock) 1:200 to 150 μg/mL in TD buffer, PBS, or colorless medium prewarmed to 37 °C (*see* **Notes 5** and **6**).
2. Remove growth medium from reporter-expressing cells and replace with D-luciferin solution.
3. Incubate at 37 °C for 10 min (*see* **Note 5**).
4. Remove the lid from the culture plate and place in plate reader or beneath the CCD camera.
5. Run a sequential imaging sequence—no filter, 510 ± 30, 610 long-pass—to obtain measurements necessary for spectral deconvolution (Table 1) and experimental analysis (Table 2) (*see* **Note 7**). The exposure times needed depend upon the intensities of the luminescence signals which are determined by reporter expression levels and imaging conditions. Other exposure settings for the IVIS imaging system are: binning—8; field of view—15; f/stop—1.

Table 1
Spectral unmixing measurements

Name	Description	Filter
Total red luminescence (R)	Luminescence from CBR-expressing cells with no filter	No filter
Total green luminescence (G)	Luminescence from CBG-expressing cells with no filter	No filter
Red luminescence in red field (Rrf)	Luminescence from CBR-expressing cells through the red filter	610 long-pass
Green luminescence in red field (Grf)	Luminescence from CBG-expressing cells through the red filter	610 long-pass
Red luminescence in green field (Rgf)	Luminescence from CBR-expressing cells through the green filter	510 ± 30
Green luminescence in green field (Ggf)	Luminescence from CBG-expressing cells through the green filter	510 ± 30

Table 2
Experimental measurements

Name	Description	Filter
Luminescence in red field (Lrf)	Luminescence from dual-color-expressing cells through the red filter	610 long-pass
Luminescence in green field (Lgf)	Luminescence from dual-color-expressing cells through the green filter	510 ± 30

6. Perform spectral deconvolution to unmix the CBR and CBG bioluminescence signals: *Manual calculation*: Enter luminescence signal values obtained from imaging into the calculations below (*see* **Notes 8** and **9**).

$$\text{CBR(Unmixed)} = \frac{\text{Lrf} - \left[\text{Lgf} \times \left(\frac{\text{Grf}}{\text{Ggf}}\right)\right]}{\left(\frac{\text{Rrf}}{\text{R}}\right) - \left(\frac{\text{Rgf}}{\text{R}}\right) \times \left(\frac{\text{Grf}}{\text{Ggf}}\right)}$$

$$\text{CBG(Unmixed)} = \frac{\text{Lgf} - \left[\text{CBR(unmixed)} \times \left(\frac{\text{Rgf}}{\text{R}}\right)\right]}{\left(\frac{\text{Ggf}}{\text{G}}\right)}$$

Proceed to **step 12**

OR

7. Perform spectral deconvolution to unmix the CBR and CBG bioluminescence signals **calculating with an ImageJ Plugin** (*see* **Note 10**) following **steps 7–11**:
8. Import 3 TIF files (open, green filter, and red filter) of experimental dual-color cells with single-color controls in the same image into ImageJ.

9. Stack images. This will merge the three TIF file windows into one window with three distinct fields (open, green, and red) that can be toggled between using the bar at the bottom of the window. Adjust image size as need.
10. Install and open ImageJ spectral unmixing plugin [11]. The spectral unmixing plug-in can be downloaded at https://imagej.nih.gov/ij/plugins/spectral-unmixing-plugin.html.
11. Draw regions-of-interest (ROIs) around single color control wells and run spectral unmixing plugin. This will generate a new unmixed image window with two distinct fields—one representing pure CBG signal and the other representing pure CBR signal.
12. Draw ROIs around experimental wells in unmixed image and measure bioluminescence signal within each ROI to obtain unmixed CBR and CBG signal values.
13. Divide the unmixed red signal by the unmixed green signal to obtain a CBR–CBG signal ratio.
14. Optional: Normalize experimentally manipulated samples to negative controls by dividing the experimentally manipulated CBR–CBG signal ratio values by control ratio values (*see* **Note 11**).

3.5 Reporter Western Blotting

1. Remove medium from cells and wash once with PBS (1.5 mL/well for 6-well plate).
2. Lyse cells directly in wells (if adherent) with 200 μL of SDS sample buffer/well for 6-well plates (scale as necessary if using different well or plate sizes).
3. Sonicate samples for 5 s with an output of five using a sonic dismembrator probe sonicator.
4. Run a portion of the sample on a 10% acrylamide gel and transfer to a PDVF membrane.
5. Probe membrane with HA antibody, which detects both CBG-TCR(WT) and CBR(TCR(PTC) protein products on the same blot. The CBG-TCR(WT) protein runs at ~75 kD and CBR-TCR(PTC) at ~60 kD. Inhibition of NMD will cause an increase in the CBR-TCR(PTC) protein level (Fig. 1b).

3.6 Reporter RT-qPCR

1. Remove medium from cells and wash once with PBS (1.5 mL/well for 6-well plate).
2. Lyse cells directly in dishes with 350 μL/well (for 6-well plate) of lysis buffer (a component of the RNA isolation kit) supplemented with β-mercaptoethanol (1%) if recommended by the manufacturer. Transfer to an RNase-free microfuge tube to store at −80 °C or proceed immediately to **step 3**.

Table 3
Master mix recipe for 10 μL reactions

Component	Volume (μL)
PowerUp SYBR green master mix (2×)	5
H_2O	2.9
Forward qPCR primer (20 μM stock)	0.3
Reverse qPCR primer (20 μM stock)	0.3

3. Isolate RNA using the RNA isolation kit according to the manufacturer's instructions.
4. Using NanoDrop obtain RNA concentrations of the samples.
5. Add 1 μg of total RNA to a PCR tube and bring volume to 13 μL with RNase-free H_2O for reverse transcription.
6. Heat RNA to 65 °C for 5 min to remove secondary structure and immediately place the samples on ice.
7. Add 7 μL mixture of oligo-dT primers (2.5 μM final concentration), random hexamers (5 μM final concentration), PrimeScript RT enzyme mix 1, and RNase-free H_2O to RNA samples in PCR tubes. Mix by pipetting.
8. Incubate samples at 37 °C for 15 min to facilitate cDNA synthesis. Terminate reactions with a 5 s incubation at 85 °C.
9. Dilute synthesized cDNA samples with RNase-free H_2O 1:8 (20 μL cDNA into 140 μL ddH_2O). Store at −20 °C or proceed to **step 10**.
10. For each primer pair, prepare a master mix of the reagents listed in Table 3. Prepare enough for the number of necessary reactions plus 10% extra; keep in mind that samples should be run in at least triplicate and no template controls should be run for each primer set. Mix thoroughly and briefly centrifuge.
11. Transfer 8.5 μL of master mix to each well on an optical plate.
12. Add 1.5 μL of cDNA samples to each well.
13. Seal wells with optical caps or film and briefly centrifuge to remove bubbles.
14. Run the qPCR using the cycling conditions below. Both fast and standard protocols require a 2 min polymerase activation step at 50 °C and a 2 min initial denaturation at 95 °C prior to cycling.
    (a) Fast (2-step) protocol: denaturation 95 °C × 1 s; annealing/extension 60 °C × 40 s; cycle number: 40.
    (b) Standard (3-step) protocol: denaturation 95 °C × 10 s; annealing 55 °C × 15 s; extension 72 °C × 40 s; cycle number: 40.

15. Calculate linear values for both CBR-TCR(PTC) and CBG-CTR(WT) mRNA expression relative to mRNA expression of controls such as GAPDH, HPRT, and β-Actin using the equation below where the ΔCt is the difference between the mean *Ct* value for the reporter gene of interest and the mean *Ct* value for the control gene.

    Linear expression value $= 1000 \times (2^{-\Delta Ct})$

16. Divide the CBR-TCR(PTC) value by CBG-TCR(WT) value to obtain a CBR–CBG signal ratio.

4 Notes

1. Cells can be infected with reporter adenoviruses ≥24 h prior to bioluminescence imaging for NMD analysis. We obtained identical results in cells infected with reporter viruses 6 days prior to imaging and cells infected 24 h prior to imaging.
2. Appropriate dilution of reporter adenoviruses will depend upon the cell lines used and the titer of the viruses and can be determined by infecting cells with various concentrations of virus and assessing reporter expression and sensitivity to known NMD inhibitors such as caffeine. For all cells lines tested, appropriate dilutions range from 1:2000 to 1:10,000 for imaging with the IVIS-50 Imaging System (Xenogen).
3. Cell lysis will generate higher bioluminescence signals. Cells can be lysed using Chroma-Glo (Promega) or an equivalent reagent according to the manufacturer's instructions.
4. Black-walled cell culture plates are preferred for bioluminescence imaging, because they minimize signal overlap between wells. 96-well plates are ideal for imaging experiments as they allow for many experimental conditions and technical replicates on the same plate.
5. Imaging can be carried out at room temperature, albeit with reduced signal intensity [12]. If imaging at room temperature, equilibrate cells to room temperature before the addition of the D-luciferin substrate.
6. The presence of phenol red during the imaging reaction reduces the amount of detectable CBG luminescence [12].
7. The filters listed in Tables 1 and 2 are recommended by Promega for use with the CBG and CBR luciferases. However, other similar filter sets with proper wavelengths cutoffs can also be used. If greater separation of CBG and CBR signal is desired, a 650 nm long-pass filter can be substituted for the 610 nm one. For imaging with an IVIS CCD camera imaging system, we routinely use a 540 ± 20 green filter for imaging

CBG activity. This filter admits a narrower range of green signal nearer the peak emission wavelength of the CBG luciferase.

8. For manual calculation method, there is no need to repeatedly image single color controls for spectral unmixing. Calibration constants for a particular combination of filter sets, imaging devices, and imaging conditions should remain constant over time. This is in contrast to the ImageJ spectral unmixing plugin, which necessitates having single-color controls present in the same image as experimental samples.
9. These calculations can be done easily by using an excel template provided by Promega found at https://www.promega.com/resources/tools/chromaluc-calculator/.
10. The ImageJ spectral unmixing requires a CCD-camera system for imaging and that single-color samples be present in the same image alongside experimental samples.
11. This normalization yields a fold-change in the CBR–CBG signal ratio of experimental cells relative to controls, which minimizes variability between plates and experiments. This step may be useful if the baseline CBR–CBG signal ratio varies between experiments while the relative change in the signal remains constant or if you wish to place specific emphasis on the fold-change in the ratio. Note that this method of normalization removes the variability between control samples from different experiments, which precludes the use of certain statistical tests.

References

1. Kurosaki T, Maquat LE (2016) Nonsense-mediated mRNA decay in humans at a glance. J Cell Sci 129(3):461–467
2. Losson R, Lacroute F (1979) Interference of nonsense mutations with eukaryotic messenger RNA stability. Proc Natl Acad Sci U S A 76 (10):5134–5137
3. Lykke-Andersen S, Jensen TH (2015) Nonsense-mediated mRNA decay: an intricate machinery that shapes transcriptomes. Nat Rev Mol Cell Biol 16(11):665–677
4. Maquat LE, Kinniburgh AJ, Rachmilewitz EA et al (1981) Unstable beta-globin mRNA in mRNA-deficient b o thalassemia. Cell 27 (3):543–553
5. Frischmeyer PA, Dietz HC (1999) Nonsense-mediated mRNA decay in health and disease. Hum Mol Genet 8(10):1893–1900
6. Hug N, Longman D, Caceres JF (2016) Mechanism and regulation of the nonsense-mediated decay pathway. Nucleic Acids Res 44 (4):1483–1495
7. Boelz S, Neu-Yilik G, Gehring NH et al (2006) A chemiluminescence-based reporter system to monitor nonsense-mediated mRNA decay. Biochem Biophys Res Commun 349(1):186–191
8. Paillusson A, Hirschi N, Vallan C et al (2005) A GFP-based reporter system to monitor nonsense-mediated mRNA decay. Nucleic Acids Res 33(6):e54
9. Nickless A, Jackson E, Marasa J et al (2014) Intracellular calcium regulates nonsense-mediated mRNA decay. Nat Med 20 (8):961–966
10. Gudikote JP, Wilkinson MF (2002) T-cell receptor sequences that elicit strong down-regulation of premature termination codon-bearing transcripts. EMBO J 21 (1–2):125–134
11. Gammon ST, Leevy WM, Gross S et al (2006) Spectral unmixing of multicolored bioluminescence emitted from heterogeneous biological sources. Anal Chem 78(5):1520–1527
12. Welsh DK, Noguchi T (2012) Cellular bioluminescence imaging. Cold Spring Harb Protoc 2012(8). https://doi.org/10.1101/pdb.top070607

INDEX

A

Ago-RIPseq 131–139
ARE binding proteins (AUBPs) 111
AU-rich element (ARE) 111, 113

B

Bioinformatics .... v, 90, 92, 107, 109, 122, 134, 137, 138, 146
Bioluminescence imaging 215–217, 219, 222
Biotinylation 18, 26, 29
BRIC-seq (5'bromo-uridine IP chase) 1–12
5-Bromouridine (BrU) 2–6, 11

C

cis-regulatory element 37, 111
CLIP-seq 2, 69

D

Dot blot 27, 30
Drosophila 56, 89, 205

E

eCLIP 111, 113, 117, 119
Electrophoresis 58, 149, 150, 152, 153, 164, 168

F

File formats
- bed 112, 117, 119, 124–126
- ENCODE narrowPeak 126
- fasta 11, 112, 116, 119, 122, 123
- fastq 6, 112, 122, 123, 137

FLAG-tag 72
Fluorescence-activated cell sorting (FACS) 84, 86, 190–192, 194–198, 200, 201
Fluorescent protein 188
Fluorescent reporter construct 78

G

Gene annotation
- AmiGO 113, 122
- biomaRt 112, 113, 117, 120, 122
- DAVID (Database for Annotation, Visualisation and Integrated Discovery) 113, 119, 120, 122
- GO (Gene Ontology) 113, 119, 122, 142, 144, 145
- MEME 113, 119, 122
- QuickGO 113, 122
- TopGO 113, 122

Genetically encoded sensor 187
GFP reporter construct 77

H

Heatmap 118
HITS-CLIP (high-throughput sequencing following UV cross-linking and immunoprecipitation) 90, 92, 112, 132

I

iCLIP (individual-nucleotide resolution CLIP) 89, 91, 94, 101, 112
Immunoprecipitation 2, 56, 60–62, 90, 93, 94, 98, 99, 112, 133, 135, 136, 143, 149, 150, 155, 175, 183
Inducible promoter
- galactose 37
- tetracycline 37, 161–163, 165, 166, 171, 172

K

Katushka fluorescent label 188, 193–195, 202

L

Lentiviral transgenesis 191

M

Mass spectrometry (MS) 56, 72, 112, 149, 150, 155
- metabolic RNA labellingradioactive nucleosides 38
- 4-thiouridine (4sU) 25–34, 38, 56, 59, 72, 73

Microarray .... v, 26, 37, 142–147, 205, 207, 209
microRNA (miRNAs) 90, 94, 111–113, 121, 131, 133, 147, 188
mRNA decay .... v, 1, 15–24, 35–52, 77, 78, 82, 86, 90, 92, 112, 119, 149, 161, 162, 213
mRNA half-lives 35–39, 46, 47, 51
Multicolored bioluminescence-based reporter system 213–224

Shireen R. Lamandé (ed.), *mRNA Decay: Methods and Protocols*, Methods in Molecular Biology, vol. 1720, https://doi.org/10.1007/978-1-4939-7540-2, © Springer Science+Business Media, LLC 2018

N

Next generation sequencing (NGS) 56, 112, 113
Nonsense-mediated mRNA decay (NMD) .... v, 161, 163, 171, 175–185, 187, 205, 213
Northern blot.............37, 162, 165, 166, 168–171, 188

O

Online databases and repositories
  DAVID (Database for Annotation, Visualisation and Integrated Discovery) 113, 119, 120, 122
  ENA (European Nucleotide Archive)........... 113, 121
  ENCODE 111–126
  Ensembl 112, 117, 120, 122, 126, 134, 138
  GENCODE..12, 116, 121
  GEO (Gene Expression Omnibus) 113, 115, 121
  Human Protein Reference Database...................... 145
  KEGG 119, 142, 144, 145
  miRBase 113, 121, 134, 137
  OMIM (Online Mendelian Inheritance in Man) 119
  PFAM 119
  SRA (Sequence Read Archive) 113, 121
  TarBase 145
  TCGS (The Cancer Genome Atlas), , UCSC genome browser 112, 113, 120, 121, 134

P

PAR-CLIP suite 55, 69–71
Phosphorylated UPF1 followed by RNA sequencing (p-UPF1 RIP-seq).................................. 175, 183
Photoactivatable ribonucleoside-enhanced crosslinking and immunoprecipitation (PAR-CLIP).......................................55–73, 112
Piwi-interacting RNAs (piRNAs)..........................89–109
Prediction tools
  ARED (AU-Rich Element-containing mRNA database) 113, 121
  AURA (the Atlas of UTR Regulatory Activity)... 113, 121
  doRiNA..69, 113, 122
  HOMER... 113, 122
  MicroT.. 144
  MiRanda 144
  Pictar.. 144
  PITA 144
  RNA22.. 144
  starBase 69, 113, 122
  TargetScan 113, 121, 141, 144

Q

Quantitative real time-polymerase chain reaction (qRT-PCR) 37, 79, 136, 137, 143, 145, 146

R

Radiolabeling...61, 62, 73, 182
Reactivation assay 205–210
Ribonucleoprotein (RNPs) complexes 55, 73, 91, 95, 100, 101, 112
RIP-seq 131–139
RNA-binding proteins (RBPs) ....v, 2, 6, 55, 69–71, 77, 90, 92, 112, 131
RNAi knockdown 161, 162
RNA-induced silencing complex (RISC) 111, 131, 132, 138, 149–158
RNA recognition element (RREs)..........................55, 56
RNA sequencing (RNAseq) .... v, 16, 21, 111

S

Saccharomyces cerevisiae............................. 35, 46, 49, 56
shRNA 79–82, 85, 176, 197
smFISH (single molecule fluorescent *in situ* hybridization) 35, 41, 47
Software 143, 165, 171, 179, 182, 207, 209
  bash 113, 114, 116
  ClueGo 144
  Cluster 145
  Conda 114
  Cytoscape... 144, 145
  Docker 113, 114
  fimo 119
  gffread.. 119
  git 113, 115
  Java TreeView 145
  Jupyter 114
  Python 113–116
  R..22, 113, 114, 117
Splicing 175, 183, 187, 189, 193, 215
Stably engineered cell lines..............80–82, 85, 187–202

T

Temperature sensitive RNAPII alleles (*rbp1-1* mutation)...36, 37
Tet-off cells 161–172
Tetracycline.................37, 161–163, 165, 166, 171, 172
Tol2 transposition 191, 193, 195–197
Transcription inhibitor
  1,10-phenantroline 37
  thiolutin 37

U

3′-Untranslated region (3′-UTR) 77–87, 90, 111, 119, 132, 146, 175, 176, 187–189, 213

W

Western blot 61, 73, 101, 176, 180, 215–217, 219

Y

Yeast 16, 17, 35, 79, 93, 97, 105, 206–208, 210, 213